About Island Press

Island Press, a nonprofit organization, publishes, markets, and distributes the most advanced thinking on the conservation of our natural resources—books about soil, land, water, forests, wildlife, and hazardous and toxic wastes. These books are practical tools used by public officials, business and industry leaders, natural resource managers, and concerned citizens working to solve both local and global resource problems.

Founded in 1978, Island Press reorganized in 1984 to meet the increasing demand for substantive books on all resource-related issues. Island Press publishes and distributes under its own imprint and offers these services to other nonprofit organizations.

Support for Island Press is provided by Geraldine R. Dodge Foundation, The Energy Foundation, The Charles Engelhard Foundation, The Ford Foundation, Glen Eagles Foundation, The George Gund Foundation, William and Flora Hewlett Foundation, The Joyce Foundation, The John D. and Catherine T. MacArthur Foundation, The Andrew W. Mellon Foundation, The Joyce Mertz-Gilmore Foundation, The New-Land Foundation, The J. N. Pew, Jr. Charitable Trust, Alida Rockefeller, The Rockefeller Brothers Fund, The Rockefeller Foundation, The Tides Foundation, and individual donors.

About Conservation International

Conservation International (CI) is a private, nonprofit organization dedicated to the conservation of tropical and temperate ecosystems and species that rely on these habitats for their survival.

CI's mission is to help develop the capacity to sustain biological diversity and the ecological processes that support life on earth. CI works with the people who live in tropical and temperate ecosystems, and with private organizations and government agencies, to assist in building sustainable economies that nourish and protect the land. CI has programs in Central and South America, Africa, Asia and North America.

For more information on CI, contact: Conservation International, 1015 18th Street, NW, Suite 1000, Washington, D.C. 20036, Tel: (202) 429–5660; Fax: (202) 887–5188.

Sustainable Harvest *and* Marketing *of* Rain Forest Products

Sustainable Harvest *and* Marketing *of* Rain Forest Products

Edited by

Mark Plotkin and Lisa Famolare

CONSERVATION INTERNATIONAL

ISLAND PRESS

Washington, D.C. □ *Covelo, California*

Library of Congress Cataloging-in-Publication Data

Sustainable harvest and marketing of rain forest products / edited by
Mark J. Plotkin and Lisa M. Famolare.
p.   cm.
Proceedings of a meeting sponsored by Conservation International and the Asociación Nacional para la Conservación de la Naturaleza, held in Panama City on June 20–21, 1991.
Includes bibliographical references and index.
ISBN 1-55963-169-4 (acid-free paper).—ISBN 1-55963-168-6 (pbk.)
1. Sustainable forestry—Tropics—Congresses. 2. Rain forests—economic aspects—Congresses. 3. Forest products—Tropics—Congresses. 4. Forest products—Marketing—Congresses. 5. Ethnobotany—Tropics—Congresses. 6. Botany, Medical—Tropics—Congresses. 7. Palms—Economic aspects—Congresses.   I. Plotkin, Mark J.   II. Famolare, Lisa M.   III. Conservation International.   IV. Asociación Nacional para la Conservación de la Naturaleza.
SD387.S87S86   1992
338.1'74987'0913—dc20   91-43278
                                                              CIP

Printed on recycled, acid-free paper

Manufactured in the United States of America

10 9 8 7 6 5 4 3 2 1

To the memory of Charles Waterton

Contents

Preface

Of all the environmental problems we face, none is more pressing or more irreversible than species extinction. And nowhere in the world is the problem of species extinction more serious than in the tropical rain forests.

But environmentalists are often accused of being more interested in protecting the plants and animals of the rain forests than the people who live there. This represents a particularly serious accusation in the tropics, where the burgeoning rates of population growth are sometimes the driving force behind deforestation. Ironically, however, those who accuse conservationists of being "anti-people" do not appear to have a working model that demonstrates how ecologically unsound development practices benefit the poor. Few countries in the Neotropics have escaped the twin evils of rampant deforestation and increased poverty during the last 20 years.

There is a growing realization in the environmental community that if conservation projects are to succeed over the long term, local peoples must materially benefit from these efforts. Too often, industrial forestry in the tropics has had terrible social and environmental costs. But unlike wood, nontimber forest products (fruits, fibers, medicines, and so forth) can often be harvested without *any* damage to the ecosystem. This labor-intensive, capital-extensive approach is well suited to local conditions in many tropical countries. If we are to manage the forest as a renewable resource, we must pay increased attention to the potential of nontimber products.

Nevertheless, several serious questions exist concerning the best way to bring nontimber forest products to local, national, and international markets: What species offer the greatest promise? What levels of harvest are sustainable? How can native peoples best be compensated for their knowledge and their efforts?

In order to answer these and other questions, Conservation International (CI) and the Asociación Nacional para la Conservación de la Naturaleza (ANCON) invited experts from around the world to a meeting held in Panama City, Panama on June 20 to 21, 1991. The conference was one of the first in which indigenous peoples and conservationists, academic researchers, and industrialists from Western societies met as equals to share experiences and ideas. This book is based on the proceeds of that meeting. But it is more than a collection of loosely related essays; it contains state-of-the-art ideas and results from the leading thinkers (and doers!) working at the cutting edge of conservation and development.

The book is divided into six sections. The first section focuses on the methodologies for collecting and utilizing ethnobotanical data. Schultes, Joly, and Lagos-Witte focus on collection and organization of indigenous knowledge. Castillo, Grenand, and Mayr provide case studies (from Panama, French Guiana, and Colombia, respectively) as to how this information can be utilized. Posey concludes the section with a thought-provoking essay on how indigenous peoples can best be compensated for the use of their ethnobotanical data demonstrating the economic potential of nontimber forest products.

Duke begins the second section with a detailed analysis of the many nontimber commodities already sold in international markets. La Rotta takes a different tack, providing detailed overviews of two promising families (Lecythidaceae and Myristicaceae). DeFilipps, Brack, Toledo et al., Reining and Heinzman, Salick, and Gentry discuss the Guianas, Peru, Mexico, Guatemala, Nicaragua, and the Andean countries. Finally, Nabhan provides an in-depth comparison of products from the arid tropics.

Because palms figure so prominently in many sustainable development projects, section three critically examines some of the more promising species. May and Lescure et al. provide case studies on a Brazilian oil palm and fiber palm, while Bernal and Pedersen give excellent overviews of Colombia and Ecuador. Rioja and Añez conclude this section with chapters on an economic development project in the Bolivian lowlands.

While medicinal plants may prove extraordinarily lucrative—a single species may be worth hundreds of millions of dollars—development of new therapeutic compounds is extraordinarily difficult. The reasons for the difficulty are outlined in section four. Daly gives an update on an ambitious effort to find new botanical treatments for cancer. King discusses the problems of and opportunities for bringing new medicines into the international marketplace. Gupta et al. provide the results of an ethnopharmacological study of the Kuna Indians, and Martin estimates the local value of medicinal plants in Mexican markets.

Section five consists of several important case studies of nonmedicinal products in the marketplace. Mori focuses on Brazil nuts, while Pendelton, Ziffer, and Hidalgo discuss fibers and vegetable ivory. Niekisch critically evaluates the "green marketing" boom in Western Europe, while Moran looks at the impact of U.S. policy on American markets for nontimber forest products. Clay concludes with an insightful analysis of several successful initiatives to bring new rain forest products into the American marketplace.

When we set out to organize this conference, we intended to pull together a wide variety of information on the potential utility of rain forest plants and organize it in a manner both relevant and interesting to conservationists, botanists, anthropologists, economists, and the general public.

But the most significant consequences of the conference came at its conclusion. Together the attendees set to work at drawing up a set of guidelines for how new products can be brought to market in an ethical, ecologically and sociologically sustainable manner. We feel confident that these guidelines (see "Conclusions and Recommendations") can serve as a benchmark against which all future efforts must be measured.

Acknowledgments

This book is based on papers presented at the conference on "The Sustainable Harvest and Marketing of Rain Forest Products," held in Panama City, Panama on June 20 to 21, 1991. This conference was coordinated by Conservation International (CI), a Washington, D.C.–based nonprofit international environmental organization, with help from the Asociación Nacional para la Conservación de la Naturaleza (ANCON), Panama.

This conference was made possible by grants from The Rockefeller Foundation, The Homeland Foundation, The Overbrook Foundation, and The Barker Foundation. We would like to thank these organizations for their generous support.

At this time we would like to thank those who participated in this conference but were unable to contribute material in time for this book. They are: Louis Poveda, Hernan Segura, and Heraclio Herrera.

A special thanks to Edward C. Wolf for his thoughts and words, and Joe Ingram of Island Press. Without his enthusiasm and support this book would have taken years to publish.

Sustainable Harvest *and* Marketing *of* Rain Forest Products

Introduction

In 1812, the eccentric naturalist-explorer Charles Waterton set out from Georgetown, Guyana, accompanied by four Amerindians and one Afro-Guyanese slave. Waterton's quest—which took him hundreds of miles through uncharted jungles and into Brazil—was in search of curare. The Englishman found the plant used to make the arrow poison and brought it back to England, where he shared it with leading figures of the medical establishment. Their experiments with this rain forest product led to the use of curare as a muscle relaxant in abdominal surgery, a use that continues to the present day.

Fourteen years after his extraordinary journey, Waterton wrote his classic *Wanderings in South America,* a book that has never gone out of print. His vivid accounts of the plants, animals, and peoples of these rain forests have inspired generations of schoolchldren to choose lives devoted to tropical exploration and conservation. One of the lines from the *Wanderings* still holds true: "No doubt there is many a balsam and many a root yet to be discovered [in these forests], and many a resin, gum, and oil yet unnoticed." The relevance of Waterton's work to today's world is twofold: over 160 years ago, he realized that the most valuable aspect of the rain forest lies in the vegetable (rather than the mineral) wealth; and that we still have much to learn—especially from the forest peoples themselves.

In 1991, Conservation International decided to organize a conference that would be based on concepts developed by Waterton. Representatives of more than 30 institutions in 12 countries and 4 indigenous nations gathered in Panama City, Panama to present original research and exchange views on the sustainable harvest and marketing of plant products from tropical rain forests.

The objectives of the meeting were to clarify the ecological, cultural, ethical, and economic obstacles that must be addressed in order to realize the potential of rain forest products to contribute to conservation and sustainable community development in tropical forests, to explore promising opportunities for product development, and to foster a partnership among conservationists, indigenous peoples' groups, academic researchers, and entrepreneurs dedicated to the development of rain forest products.

The workshop was prompted by alarm at the continuing destruction of tropical rain forests—the most species-rich ecosystems known on Earth—and the parallel erosion of indigenous forest-dependent cultures; by recognition that conventional approaches to conservation are no longer adequate to meet the scale and rate of destruction unfolding in tropical countries; and by encouraging evidence that the sustainable harvest of nontimber forest products is technically possible, financially attractive, and compatible with the linked goals of ecosystem conservation, cultural survival, and viable community development.

I

Conserving
Ethnobotanical Information

1

Ethnobotany and Technology in the Northwest Amazon: A Partnership

RICHARD E. SCHULTES
Botanical Museum of Harvard University

Technology and ethnobotany—what relationship have they with each other?

We must first define what is meant by *ethnobotany*. Normally, *ethnobotany* today means the complete registration of the uses of and concepts about plant life in primitive societies. It is an ancient field, but only in the last century has it assumed the status of a distinct research branch of the natural sciences. It is interdisciplinary, comprising aspects of botany, anthropology, archaeology, plant chemistry, pharmacology, history, geography, and sundry other tangential fields of the sciences and arts.

The earliest human beings must have been ethnobotanists: they were forced to classify the plants of their ambient vegetation; distinguishing those of definite use, those without any known use, and those that were toxic, sometimes even fatal. Eventually, they had to develop techniques to utilize many of the species, including the poisonous ones.

And what really is *technology*? Coming from Greek roots meaning systematic treatment, it is today defined in our dictionaries as "the science of systematic knowledge of the industrial arts." A more inclusive anthropologically oriented definition holds that technology is a "body of knowledge available to civilization that is of use in fashioning instruments, practicing manual arts and skills and extracting or collecting materials." Now, I do not want to set myself up as a glossological expert, but I have often felt that an even broader meaning is desirable, and I propose here to interpret the term as the human manipulation of natural materials, phenomena, or substances.

Consequently, I would like to recognize two kinds of technology.

One concerns the technological treatment by native peoples of indigenous ethnobotanical knowledge; we might call it *aboriginal technology*. The second is the advanced or complex technology of modern industry and science. And I hope to illustrate how modern technology has utilized and can often even extend some of the discoveries made in native societies over the centuries and which comprise present day aboriginal technology.

The perspicacity of the Amazonian Indians is unbelievable, especially in their relationship with plants and animals. They are literally masters of their ambient vegetation. Their knowledge of the properties of the plants of their environment is deep. Amassed over millennia, it has been passed on orally from one generation to the next. Unfortunately, today it is in grave danger of being lost as a result of acculturation and "westernization" that is increasing in most parts of the world where native peoples are exposed to expanding commercial activities, growing missionary persistence, tourism, road and dam building, warfare, and other pressures. Much of this precious knowledge is disappearing faster even than the trees in many regions where forest devastation is uncontrolled and rife. Its loss will be disastrous for the progress of humanity as a whole.

Those of us working for protection of the environment have, in recent years, drawn attention to the urgent need of ethnobotanical conservation.

Of the very many available examples of clever technology practiced by the Indians of the northwest Amazon, I shall mention only several that illustrate, I believe, their acute inventiveness: 1) in the preparation of curares or arrow poisons; 2) the use of one of their principal sacred hallucinogens; 3) how they free their major source of carbohydrate food from a deadly poison; 4) the several complex ways in which resin-like exudation of a tree is utilized; and 5) their use of certain plants in fishing.

Primitive societies on every continent have learned to employ plant or animal poisons on their darts and arrows. And on nearly all continents, natives have invented blowguns or blowpipes from which they can shoot their poisoned darts with incredible accuracy.

Indians of the northwest Amazon prepare the greatest diversity of curares with the most complex technologies. More than 75 species of plants in many genera and families are employed in the Colombian Amazon alone. These Indians elaborate many different types of curare, according to the size, kind, and habits of the animals that they want to hunt. It is probably no exaggeration to say that every tribe and often every medicine man may have their own curare formulas. An arrow poison may comprise several or many different plants—frequently using as many as 15 ingredients; but some curares are made from a single species. We are still discovering curare plants that have never been reported.

Usually, one or two of the plant ingredients of curare contain toxic compounds and are physiologically effective; often, however, there are many additives that are free of toxic principles but which may alter the properties of the bioactive plants—an effect scientifically known as *synergism*.

It is difficult to understand how the Indians could discover, in the Amazonian flora of 80,000 species, so many plants that, properly prepared, provide deadly arrow poisons. Not only is the discovery of this number of toxic plants in such a rich flora extraordinary; the often complex techniques and combinations required in making the various poisons bear witness to the uncannily inventive skill of these natives.

Modern medical technology has acquired a valuable drug from one of the curares of Colombia, Ecuador, and Peru. The active principle resides in the bark of a gigantic forest liana known botanically as *Chondrodendron tomentosum*. In the 1930s, the active constituent in this plant—an alkaloid called tubocurarine—became important in modern surgery as a skeletal muscle relaxant for treating myasthenia (abnormal muscular fatigue) and certain neuropsychiatric conditions.

In many primitive societies, there is no concept of organically caused illness or death, which they believe results from the interference from malevolent spirits who dwell in supernatural realms. As a consequence, medicine men often diagnose and even treat disease by contacting these mythical beings through visual and auditory hallucinations. The hallucinogens, with such potent psychic effects, are considered to be sacred plants.

The most widely used hallucinogen in the northwest Amazon—variously known as ayahuasca, caapi, or yajé—is a drink prepared from the bark of an extensive wild liana botanically called *Banisteriopsis caapi*. It is usually ceremonially used and has several alkaloids of the beta-carboline type that induce visual hallucinations without interfering with muscular activity; in fact, dancing is often part of the ceremonies. Frequently, medicine men cultivate the vine.

In some tribes, other plants—often themselves toxic—may be added to strengthen or lengthen the intoxication. The most commonly employed additives are the leaves of two plants: a shrub called chacruna (*Psychotria viridis*) or a vine known as oco-yajé (*Diplopterys cabrerana*). These leaves contain other intoxicating alkaloids of the tryptamine type. When orally ingested, however, tryptamines are inactive, unless accompanied by a chemical compound known as a monoamine oxidase inhibitor capable of activating tryptamines taken orally. The leaves of these two additives contain tryptamines that do enhance and lengthen the hallucinogenic effects. Worth noting is that the tryptamines in these leaves are activated by the beta-carboline alkaloids in the *Banisteriopsis*. Without the presence of these alkaloids the tryptamines would be destroyed in the stomach. Here we have another example of Indian technological knowledge of plant properties, although they do not understand the chemistry involved.

While modern medicine has not yet found use for these several alkaloids, we still may marvel at the technology of the Indians who, in a massive vegetation of 80,000 species, could discover certain plants containing monoamine oxidase inhibitors or learn the effects of these invisible chemical substances.

A third example of Indian technology concerns detoxification of the cassava plant (*Manihot esculenta*), their principal carbohydrate food.

The strains of the cassava plant that are cultivated in the Amazon contain a cyanide in the root. An ingenious method of removing this toxic substance has been devised—an excellent example of aboriginal technology.

The plant has a three- to four-foot root capable of reaching nutrients in the usually poor soils of this humid region of rain forest, where cereal grains with superficial roots cannot easily be cultivated. The root is first grated, and the starchy mash is soaked overnight in water in a dug-out canoe. The next day, the mash is packed into a long woven cylinder or press called a tipi-tipi. When it is filled to capacity, the tipi-tipi is hung on a protruding branch of a tree. A pole is attached through a loop in the lower end of the apparatus, and one or two women sit on the pole. The tipi-tipi is thus squeezed, expelling most of the cyanide-containing liquid. The still humid mash is then removed and heated on a flat stone over a fire. The dried result is a coarse meal from which unleavened bread can be prepared.

A fourth example of ingenious technological discovery among aborigines of tropical America pertains to three uses of a red resinous exudate of the inner bark of several species of trees in the genus *Virola* of the nutmeg family.

From this resinous substance, many Indian tribes in the northwestern Amazon prepare a hallucinogenic snuff of extreme importance to these people. The bark is stripped from the tree early in the morning, and the resin is collected, heated until it is a thick paste, sun-dried, and pulverized; the resulting fine powder is the snuff known among the Waika as epena or nyakwana. It has an extremely high concentration of potent tryptamine alkaloids (up to 11 percent). Frequently, the powdered, highly aromatic leaves of a small shrub known as màshihàra (*Justicia pectoralis*) are added. Mashihara is itself hallucinogenic and is often used alone as intoxicating snuff.

Epena snuff creates first a feeling of bravado, and the men leave the large round-house to fight off evil spirits (the hikuri) about to harm the tribe. Following the expenditure of so much energy, they return to the house and take more snuff and engage in strange movements, mimicking animals. Following a period of lethargy, they sink into a deep sleep disturbed by frightening visual hallucinations.

The second use of the resin is as a weak arrow poison. It is smeared on the blow-gun darts and gently heated to solidity.

The third use of the resin and—from the viewpoint of our modern technology the most important—is medicinal. Fungal skin infections are extremely common in the wet tropics. Our own pharmacopoeias have no real cures for most fungal skin infections: athlete's foot, jock itch, and so forth. We have merely suppressants. Over wide areas of the wet tropics of South America, natives in Colombia, Ecuador, and Suriname treat these conditions by coating the infected areas daily for 10 or 15 days with the resin of *Virola*. The redness in the infected areas disappears: whether this be a cure or only suppression is still not known. Recent pharmaceutical tests, however, have given very promising results. We may well be on the way to discovering from aboriginal technology a new and much needed agent in treating these widespread and bothersome infections.

Another example of aboriginal technological discovery concerns the use of plants in fishing. Some plants contain chemicals that interfere with gill function, causing the fish to come to the surface for oxygen. There they are caught by hand by an Indian sitting in a canoe. The principal Amazonian fish poison is from the bark of *Lonchocarpus,* a vine of the bean family.

We do not want to poison our lakes, but chemists have found that the active principle is a ketone called rotenone. Modern technology has bent rotenone to our use as an agricultural pesticide which, unlike DDT, is rapidly biodegradable, breaking down in several days, thus not entering the human body through vegetable and animal foods.

It may be pertinent to consider how best modern technology can profit from aboriginal technology by working together in the field.

Throughout my 47 years of plant exploration in the Colombian Amazon, I moved about in a canoe with native paddlers. In 1977, however, my colleague, Professor Holmstedt (the Karolinska Institute in Stockholm) and I had the good fortune of directing ethnopharmacological research *in situ* amongst the Witoto and Bora Indians in the Ampiyacu and Yaguasyacu rivers in the Peruvian Amazon with an international group of 14 specialists: botanists, zoologists, chemists, pharmacologists and biochemists from seven countries.

We were based on a fully equipped floating laboratory, the RV Alpha Helix, a ship attached to the Scripps Oceanographic Institute and financed by the National Science Foundation. A research ship used mainly for oceanographic study, it was sent to the Amazon in 1977 for a year, which was divided into six two-month phases of various scientific projects. While each scientist on our Phase VII had his or her own specific interest, much of the research was cooperatively carried out and centered on investigations of the local Indian drug lore. Some of the projects could be finished on board, while material for further research at home was collected. Our analyses of bioactive plants were highly successful. The available sophisticated modern equipment permitted analytical studies of the chemical constituents of plants to be carried out on extracts of a few leaves collected by the botanists with Indians in the jungle an hour or so before arrival on board ship. It was not necessary to collect 50 or more pounds of leaves, dry them, possibly altering their chemical constituency, and ship the material to far distant laboratories in foreign countries for eventual study. The botanists were able to collect 960 plants representing 3500 specimens. A total of 150 species were ethnobotanically and chemically studied during the two-month phase.

Much new information was acquired, primarily concerning the chemistry of a number of species employed by these Indians as medicines or hallucinogens. Among numerous other significant research projects were ethnobotanical, chemical, and biological studies of the coca plant, source of cocaine (*Erythroxylon coca*), the pulverized leaves of which Indians of the northwest Amazon employ as a narcotic, chewing the powder for its stimulant effects. It was possible to carry out on board monitoring and quantitation of the peak heights and retention time of deuterated cocaine, following chewing of the powdered coca leaves by members of the scientific crew who, unlike the natives, were not habituated to the use of coca. This very significant study would have been impossible without freshly gathered leaves.

Such extremely productive ethnobotanical, biological, and chemical research could be carried out only with the most modern and sophisticated technological facilities and on location in the rain forest where we had the cooperation of the Indians who shared with the scientists their knowledge and aboriginal technology.

There are innumerable examples of ethnobotanical technology

among peoples in primitive societies around the world that have been, and others that could be, of incalculable value to modern technology.

Even in the United States there are numerous examples. The beautiful Indian tobacco (*Lobelia inflate*) used as a tobacco substitute by the Indians contains lobeline, now commercially employed as a smoking deterrent. The Mayapple (*Podophyllum peltatum*) has a toxic resin formerly used by Indians in New England to remove warts; its cytotoxic action has led to its modern use in treating soft venereal warts, which formerly required surgery. The cardiotonic strophanthine comes from seeds of a species of *Strophanthus* employed by African natives as the source of their arrow poison. The alkaloid resperine, now a valuable hypertensive and tranquilizer from the root of a small shrub (*Rauwolfia serpentina*), has been used in India for more than a thousand years in folk medicine for snake bite and insanity. Natives of Madagascar value the periwinkle (*Catharanthus roseus*) as an oral hypoglycemic (reduction of the sugar content of the blood) agent: it has 55 alkaloids, one of which has given modern medicine an extremely successful agent in the treatment of leukemia. We could name many more examples that have led modern technologists to discoveries. But there are many more in the world that have as yet never been tapped by science or industry.

It is my hope that the few examples that I have cited may serve to emphasize the urgency of ethnobotanical conservation and to stress the great value of aboriginal technology and its study to modern technological advances in many fields of science and industry for the good of all humanity. Time, however, is of the essence, for many aspects of aboriginal technology will disappear during the next half century.

Note: This lecture was delivered on the occasion of the author's receiving the Lindbergh Award for contributions of balancing technological advancement and environmental preservation. It was given at the Explorers' Club in New York on May 7, 1991.

2

The Secrecy of Indian Ethnobotanical Data Banks

Luz Graciela Joly
Universidad Nacional de Panama

Numerous ethnobotanical researchers have noted Amerindian ethnopharmacopeias are often closely guarded secrets. Scientists must often spend considerable time gaining the trust of indigenous colleagues before they gain access to this information. After five centuries of European colonization, Amerindians have derived relatively few socioeconomic benefits from plants and animals that they first domesticated and used that have subsequently been commercialized by Westerners. It is no wonder then that confidence-building between scientists and indigenous peoples may be a lengthy process.

The retrieval of ethnobotanical information from Amerindian data banks should be based on ethical behavior according to the following precepts: first, ethnobotanical information is an economic resource that has a price in Amerindian medicinal and other social/cultural practices in the same way that information is an economic resource in Western science. Amerindians who are knowledgeable in ethnobotanical uses ordinarily charge their own people for their services or for the release of the information. Therefore, researchers and the institutions that fund research should also pay for this information. Furthermore, they should pay as much as it costs for Amerindians and then pay much more. If the information is used to create pharmaceutical or other products, a percentage of the profits should also be paid to the Amerindian population. These payments should be made in a way that will benefit not just the one or two informants but the whole Amerindian group among whom that information is a social/cultural heritage. Many Indian groups are losing their traditional territories. Payment for information from Amerindian data banks could be made to gain legal title to their traditional lands. Monies could also be used to fund bicultural and bilingual educational and health programs.

Many Amerindians have made tremendous sacrifices to become formally educated in Western scientific methods. There are now many professional Amerindians and, whenever possible, these people should be included as part of the research teams and should be given full credit as coauthors in publications—not merely acknowledgments at the end of an article as so often happens. If there are no professional Amerindians available to participate in the research, an effort should be made to grant scholarships to members of these Amerindian groups so that they also may have the opportunity to expand the use and knowledge of their own data banks. Likewise, those Amerindians who supply the information (Shaman, curanderos, and so forth) and other informants should be given credit and their names be recognized historically. For example, plant species should be named after them.

Ethical principles outlined in this chapter create confidence, dignity, and respect for fellow human beings who continue to be discriminated against culturally, economically, and educationally, and will ensure that ethnobotanical studies carried out along these lines will benefit everyone—not just the scientists themselves.

3

Five Hundred Years of Tropical Jungle: Indigenous Heritage for the Benefit of Humanity

GEODISIO CASTILLO
Fundación Dobbo Yala, Panama

Without a doubt, the tropical rain forest is part of our lives. Here, as in no other place on our planet, we feel and find ourselves to be better off because of a great goodness that our friends the trees give us. If we knew the goodness of their cover and support, of the air we breathe, of the many creatures that partake of their fruits. . . . But all this biological exuberance is developed within a fragile ecosystem over an extremely poor, often toxic soil, which is located beneath a slender layer of soil. So, we believe in the type of "development" that has been used to justify the creation of an environmental problem that has destroyed the trees. We human beings can no longer continue to rape our own Mother Earth, to plunder her inner spirit, which permeates all things.

This negative action has also influenced our indigenous cultures, subtly imposed by foreign education and religion, economic contacts, and the colonization of our territories considered to be "empty." Since 1492, these ethnicidal actions have caused the disappearance of much indigenous ecological knowledge. We wish to make clear that we do not subscribe to the myth that we indigenous peoples never destroy natural resources. We do; however, we wish to underscore that we are indeed conservationists, thanks to our ecologically oriented culture and our detailed knowledge of the environment.

Many barren landscapes have resulted from ill-planned development. This immense process of incessant changes will negatively affect the direction of human civilization. Too often, this type of economic activity increases at the expense of others, such as other species that go extinct, or the climate that continues to degrade.

We have just recently heightened the environmental awareness in the context of the political process. Now we realize that some of the greatest riches of our planet, plant and animal species, live in the jungles and natural forests as part of the natural services furnished by the ecosystems. Because the existence of this natural capital is indispensable, its conservation is important not only for sustainable development, but also for our very survival.

In our country, many of these natural resources are located either in reserves or indigenous reservations. Therefore, the indigenous territories are in no way "empty" or "vacant"; nor do the resources slow down national development. Quite the contrary, they safeguard the natural and cultural heritage of the country.

In reviewing the existing resources in the indigenous jungle of our country, we find many tree species, such as udirbi (*Colpothrinax cokii*), sapigarda (*Simaba polyphylla*), Spanish cedar (*Cedrela odorata*), spiny cedar (*Bombacopsis quinatum*) and other genera such as *Brosium, Rheedia, Sanblasia,* and *Plowmania,* among others. During a recent exploration to Cerro Tacarkuna in southern Panama, 292 known plant species were found, of which 24 are endemic in the higher part of the mountain, above 1,500 meters. Nevertheless, although many species may be unknown to science, they are not new to the indigenous inhabitants.

We should note as well that, according to studies undertaken by PEMASKY and the Smithsonian Tropical Research Institute, 100 species of mammals, 440 bird species, and some 30 amphibious species were found in the Kuna Yala Reservation. Microfauna is even more numerous than the vertebrates, particularly arachnids and insects, among which many butterflies are particulary numerous. A distinct trait of the area is the existence of many coralline islands. Fish abound there, as do shrimp, lobster, and four sea turtle species.

Thus, as the Dobbo Yala Foundation understands, it is important for every citizen to respect our Mother Earth. We must find strategies that allow our energies to be channelled so that they act for her benefit. Indigenous people must participate in official development plans—from their design, through their execution and evaluation—in order to guarantee the existence of natural resources. We find excellent examples in the agroforestry techniques of our peoples, in traditional alternative crops, and in cyclical agriculture. Several ecologists have recently learned major lessons by studying our indigenous techniques of enriching the jungle with vegetation that, until now, had been thought to be wild.

We also believe it is important to promote planning and development in unspoiled areas on the basis of our indigenous philosophical principles. It is important to analyze the impact of outside funding on our indigenous culture to determine whether is it positive or negative. We must analyze the adverse consequences of the Western development models on ecology and the environment. We must also critically examine emerging technologies such as biotechnology (genetic engineering, specifically). In reality, these biological alternatives may be technological efforts to circumvent biological alternatives that have been proven ecologically safe. What is being attempted? Possibly the control by the countries of the North over the agriculture and genetic resources of the Third World, to further ruin our already unstable economies by depriving our cultures of development by forcing our agricultural systems to compete with genetically manipulated substitutes that may ultimately endanger the biodiversity of the planet by sending forth genetically manipulated organisms.

Environmental management must be autochthonous. Therefore, to avoid wasting everything, new economic and technological models and progressive advances must come forth in concert with ecological and cultural advances in order to manage and make good use of the biosphere, in relation to the quality of life.

Thus, in order to dignify our Mother Earth and instill humans with humanity, it is imperative to consider conservation as a key instrument for rural and cultural development; it clearly depends on both our dignity and our education. Environmental education, formal and extracurricular, constitutes the most appropriate means of ensuring citizen participation and of popularizing official scientific knowledge and techniques, thereby finding common ground with the indigenous culture.

The country's biological and cultural diversity are of intangible worth in that all species have the right to exist. When managed appropriately, biodiversity lays the foundations for recovery, sustainable development, and a solid national economy. Important political, economic, and social benefits may be achieved by integrating biodiversity, the recognition of a multicultural Panamanian state, and the legal recognition of the territories and the self-determination of the indigenous peoples.

This task of integration will lead to the best possible management of natural wealth and will guarantee that indigenous and rural wisdom is not stunted; this will contribute appreciably to lessening human needs and ensuring that the development of sovereign peoples can become a reality and not the nostalgia of tomorrow.

Without the support of this population—the poor rural farmers and indigenous people—there will be no conservation policy for biodiversity that can be successful. Nor will any development policy be guaranteed that cannot demonstrate sustainable achievements to its communities.

4

Ethnobotanical Contributions to the Tramil-Program in the Caribbean Basin: The Case of Honduras

Sonia Lagos-Witte
Universidad Nacional Autónoma de Honduras

Based on the concept that ethnobotany is an empirical science that has produced spectacular results in the understanding of plants and nature, the National Autonomous University of Honduras (UNAH) through its Research Division (DICU) and its Department of Biology has, for a number of years, encouraged ethnobotanical studies with the objective of strengthening this empirical science through a systematic documentation of those plants used by the local population for medicinal, nutritional, and industrial purposes.

Compared with the overwhelming mestizo population in Honduras, the existing indigenous cultures are relatively small in size (Bueso, 1987). The Chortí Indians are found in the west (Copán and Ocotepeque) and are linguistically and culturally descendants of the Mayas. In the eastern part of the country are the Sumos, the Miskitos (Mosquitia), and the Pech (Payas). The Jicaques are concentrated in the region of Yoro and La Montaña de la Flor in north and central Honduras, and the Lencas are dispersed throughout the whole central region.

In addition to these indigenous minorities is a Black Carib population known as Garifunas, whose ancestors came from the Caribbean island of San Vicente during the seventeenth century. The Garifunas are the product of a cultural mix of three elements—the Arawak and Carib Indians from the north coast of South America and the black slaves who were survivors from shipwrecks or who fled from the other Caribbean islands (McCanley, 1981).

Although the other 90 percent of the population of Honduras are mestizos, much of the ethnopharmacopeia of the country is based on plants originally used by the Indians and the Garifunas (Lagos-Witte, 1987).

Many studies of the ethnic groups of Honduras focused on the anthropological aspects of their cultures. Consequently, very few reports describe the utilization of plants for medicinal purposes. Some of these studies that do feature ethnobotanical information include Conzemius (1932), Coelho (1955), McCanley (1981), Velásquez (1980 & 1987), Cruz Sandoval (1984), Matamoros and Ardón (1984), and Balick (1990).

In 1985 a number of ethnobotanical studies were initiated with the following objectives: 1) to contribute to the rescue and preservation of traditional uses of medicinal and nutritional plants as utilized by the rural population of Honduras; 2) to produce a comparative overview of the plants of Honduras; and 3) to evaluate the popular uses of the most significant plants according to pharmacological and chemical criteria.

The information is based on a questionnaire designed according to the specific issues of each study. Three standard questionnaires were used: 1) plant questionnaires based on the specific uses of a single plant; 2) questionnaires based on the uses of different plants within a single community; and 3) questionnaires based on the various plants used by a single "curandero."

In the survey of the medicinal plants used by the Black Caribs as part of a regional Caribbean research program known as TRAMIL (Traditional Medicine for the Islands) another questionnaire was developed which included information on: 1) general social and economic aspects; 2) educational level; 3) access to health services; 4) common illnesses or health problems; 5) utilization of plants in their home medicine; and 6) parts of the plants being used for each specific problem.

Each plant mentioned in the interviews was then collected and made into herbarium specimens. These botanical samples were then identified by the taxonomist within the project team, with the support of the National Herbarium of the University (UNAH). All the specimens were stored in the herbarium of the Laboratory for Ethnobotany and Plant Histology at the UNAH.

THE BLACK CARIBS OF HONDURAS
AND THE TRAMIL SURVEY

According to Coelho (1955), the Black Caribs known as Garifunas arrived on the north coast of Honduras around 1797, establishing settlements on the Roatan Island and along the Atlantic coast from Belize to Nicaragua.

Today there are 43 Garifuna settlements in Honduras with an estimated population of about 60,000 inhabitants (Cruz Sandoval, 1984). They maintain their own language (Coelho, 1955) and, in contrast to the other ethnic minorities in Honduras, also most of their own culture and identity.

The TRAMIL surveys were carried out throughout 1988 and 1989 among seven Black Carib communities: Corozal, Nueva Armenia, Travesia, Tornabe, Triunfo de la Cruz, Bajamar, and Cuzuna. The climate of this region is tropical, with a dry season from March through May and a rainy season from June through November. The average temperature is 26 degrees Celsius.

During the study a total of 476 interviews were performed. Eighty percent of the informants with knowledge about medicinal plants were women older than 40 years of age. Seven illnesses or health problems were found to be very common: fever, grippe, cough, headache, stomach pains, diarrhea, and asthma.

In general terms, the living conditions of the Garifunas are better than those of other ethnic minorities in the country. Although public health services are available for 70 percent of the population within the study area, 75 percent of the informants reported using medicinal plants in their home health care practices.

Of more than 200 species identified in the surveys, 34 were considered significant and have been used as the basis for pharmacological and chemical evaluation in the TRAMIL workshops (Weninger and Robineau, 1988; Robineau, 1991).

DISSEMINATION OF THE RESULTS AND COLLABORATION WITH NATIONAL NONGOVERNMENTAL ORGANIZATIONS

The results of our study have provided interesting insights into species used for medicinal purposes by the local population. This research has also served as a contribution to the revitalization of traditional medicine of Honduras and complements ongoing research activities on the ethnobotanical inventory of the national flora. It also contributes ethnobotanical data to the TRAMIL-program coordinated by ENDA-CARIBE, the aim of which is a comparative analysis on the use of medicinal plants among the Caribs in different countries in and bordering the Caribbean region. The results will eventually permit us to perform further scientific evaluation with respect to their effectiveness and/or toxicity on the medicinal plant uses in Honduras (Lagos-Witte and Robineau, 1990). Of particular ethnobotanical importance is the fact that through these studies (since 1985) the first national ethnobotanical herbarium was organized, and it now has more than 700 specimens with their respective ethnobotanical information deposited.

Another important result of our studies has been the publication of the "Popular Manual of 50 Medicinal Plants" (1989), which includes the most frequently used plants with the parts and form of their application (House, Lagos-Witte, and Torres, 1989). The manual has been cofinanced by the UNAH, CIIR (Catholic Institute for International Relations, England) and six nongovernmental organizations with the objective of introducing this information into local health programs. Several of the researchers from the Laboratory of Ethnobotany assisted in the presentation of several workshops organized by these nongovernmental organizations in order to transmit the most important research results, based on the manual, to the rural people.

During this period there were many controversial discussions because of contradictions found between the popular uses and the real effectiveness and/or toxicity of some plants. For instance, *Jatropha curcas,* a common plant used for mouth infections (latex, in local application) and intestinal parasites (seeds roasted and swallowed with water), is classified as toxic because of the content of curcin or curcasin and a poisonous resinous substance present in the whole plant, but particularly in the seeds (Weninger and Robineau, 1988; Dharma, 1987). According to Dharma (1987), *Jatropha curcas* "is used in at least 23 countries and it is found in the WHO priority list of the most used medicinal plants in the world." The workshops permit the dissemination of this type of information to a population that otherwise has no access to the results of laboratory information. Similar situations have arisen with other commonly used plants such as *Argemone mexicana, Ricinus communis, Cassia occidentalis, Nerium oleander,* and *Lantana camara.*

As a result of the TRAMIL surveys done in the Black Carib communities and the evaluation of the medicinal plants in the TRAMIL workshops, a very important dissemination program was initiated in January 1990 in collaboration with nongovernmental organizations and the communities in which the surveys were carried out.

The increasing interest in traditional medicinal plants in the Caribbean region is also reflected in multiple recent publications about ethnobotany in Central America and the Caribbean. Some of the most important include an atlas of the medicinal flora of Central America and the Caribbean by Morton (1981); ethnobotanical studies of Mexico by

Hernández Xolocotzi (1985, 1987); uses of some medicinal plants in Costa Rica by Ocampo and Maffioli (1987); ethnobotanical inventory of medicinal plants used by the Guaymi Indians in western Panama by Joly et al. (1987, 1990); an ethnobotanical survey of the medicinal flora used by the Caribs of Guatemela by Girón et al. (1989); and finally, the TRAMIL studies compiled by Weninger and Robineau (1988) in "Elements for a Caribbean Pharmacopeia" and Robineau (1991) "Hacia una Farmacopea Caribeña" as a result of the above-mentioned interinstitutional and international cooperation.

Although the number of ethnobotanical researchers is increasing, recent studies indicate that less than 1 percent of all plant species have been studied in detail with respect to their potentially useful properties (Forero, 1989). Therefore, there is still a vast treasure of "empirical traditional science" (Gómez-Pompa, 1986) to be investigated and documented before it will be lost forever because of the penetration of what we call the "modern civilization." But even if all the ethnobotanical knowledge that still exists could be documented, we should still be aware of the fact that we may be overlooking other potentially useful plants. For edible purposes, for example, Schultes (1989) estimated that only a small portion of potentially edible species have ever been entered into international commerce. The great majority of plants and their potential use remain outside our grasp. Both Prance (1977) and Forero (1989) deplore that as "the destruction of the vegetation is proceeding more rapidly than the biological inventory, probably thousands of plants will disappear without ever having been discovered."

References

Balick, M. J. 1990. Production of Coyol Wine from *Acrocomia mexicana* (Arecaceae) in Honduras. *Economic Botany* 44 (1):83–93.

Bueso, J. A. 1987. *El Subdesarollo Hondureño*. UNAH. ed. Universitaria, Honduras.

Cámbar, P. J. 1983. *Estudio de los efectos farmacológicos de algunas plantas nativas de Honduras; Normas para un trabajo interdisciplinario: Botánico, Quimico, Farmacológico*. UNAH. Tegucigalpa, Honduras.

Cámbar, P. J., et al. 1987. *Prevención de la producción de úlceras gástricas experimentales por algunos extractos acuosos de plantas*: Chenopodium ambrosioides, Passiflora edulis, Momordica charantia, Sechium edule *y otras*. Departamento Ciencias Fisiológicas, Facultad de Medicina, UNAH. Honduras.

Cámbar, P. J., et. al. 1989. *Efectos del extracto acuoso de* Annona muricata *en la prevención de la producción de úlceras gástricas según el método de Shay.* Departamento de Ciencias Fisiológicas, Facultad de Medicina, UNAH. Honduras.

Coelho, R. G. 1955. *Los Negros Caribes de Honduras.* Publ. 1981 by Editorial Guaymuras, Tegucigalpa, Honduras.

Conzemius, E. 1932. *Ethnographical Survey of the Miskito and Sumu Indians of Honduras and Nicaragua.* Smithsonian Institution, Bulletin 106.

Cruz Sandoval, F. 1984. Los Indios de Honduras y la situación actual de sus recurses naturales. *América Indígena,* XLIV (3):423–46.

Dharma, A. P. 1987. *Indonesian Medicinal Plants.* Balai Pustaka, Jakarta, Indonesia.

Forero, E. 1989. *Current State of Botany in Latin America.*

Girón, L. M., et al. 1989. *Ethnobotanical Survey of the Medicinal Flora Used by the Carib of Guatemala.* CEMAT, Guatemala.

Gómez-Pompa, A. 1986. *La Botánica Económica: Un punto de vista.* IV Congreso Latinoamericano de Botánica, Medellin, Acad. Colombiana de Ciencias, 57–63.

Hernández Xolocotzi, E. 1985. *Xolocotzia,* Tomo I, Rev. de Geografia Agrícola. Universidad de Chapingo, México.

Hernández Xolocotzi, E. 1987. *Xolocotzia,* Tomo II, Rev. de Geografia Agrícola. Universidad de Chapingo, México.

House, P., S. Lagos-Witte, and C. Torres. 1989. *Manual Popular de 50 plantas Medicinales de Honduras.* Lit. Lopez, Tegucigalpa, Honduras.

Joly, L. G., et al. 1987. Ethnobotanical Inventory of Medicinal Plants Used by the Guaymi Indians in Western Panama. Part I. *Journal of Ethnopharmacology* 20:145–71.

Joly, L. G., et al. 1990. Ethnobotanical Inventory of Medicinal Plants Used by the Guaymi Indians in Western Panama. Part II. *Journal of Ethnopharmacology* 28:191–206.

Lagos-Witte, S. 1987. *Reflexiones en torno de la medicina tradicional de Honduras.* Memorias I Seminario Mesoamericano de Etnofarmacologia. Guatemala.

Lagos-Witte, S. 1988. Estudios etnobotinicos de plantas Hondureñas. Rev. de la Universidad. UNAH. Tegucigalpa, Honduras, VI (24):86–92.

Lagos-Witte, S., and L. Robineau. 1990. Contribución etnobotánica al proyecto TRAMIL "Investigaciones Cientificas y usos populares de plantas medicinales en el Caribe." V Congreso Latinoamericano de Botánica. La Habana, Cuba.

Matamoros, D., and M. Ardón. 1984. *Panorama de los grupos etnicos de Honduras.* Memoria I Seminario de Medicina Tradicional. Tegucigalpa, Honduras.

Matamoros, D. 1987. Análisis de las Condiciones de salud del niño de 0–6 años en Honduras. Toronto, Canada. OMEP, 19 (1).

McCanley, E. 1981. *No me hables de muerte sino de parranda.* ASEPADE, Honduras.

Morton, J. F. 1981. *Atlas of Medicinal Plants of Middle America.* Springfield: Charles C. Thomas.

Ocampo, R. A., and A. Maffioli. 1987. *El uso de algunas plantas medicinales en Costa Rica*. Vol 1.

Prance, G. T. 1977. Floristic Inventory of the Tropics: Where Do We Stand? *Annals of the Missouri Botanical Garden*. 64:659–84.

Robineau, L., ed. 1991. *Hacia una Farmacopea Caribeñia*. Seminario TRAMIL 4. Tela, Honduras. Nov. 1989.

Salud en Cifras. 1983–1986. *Dirección de planificación*. Departamento Estadistica de Salud Honduras.

Schultes, R. E. 1989. El folklore botánico y la conservación de los recurses naturales. *Fundación Peruana por la Conservación de la Naturaleza* 4:3–22. Lima, Peru.

Velásquez, R. 1980. El chamanismo Misquito de Honduras. *Yaxkin*, III (4)273–310.

Velásquez, R. 1987. *Chamanismo mito y religión en cuatro naciones étnicas de America aborigen*. Caracas.

Weninger, B., and L. Robineau, ed. 1988. *Elementos para una Farmacopea Caribeña*. Informe TRAMIL 3. La Habana, Cuba.

5

The Use and Cultural Significance of the Secondary Forest Among the Wayapi Indians

PIERRE GRENAND
ORSTOM, Paris

Because so much of the world's tropical forest has already been disturbed, any attempt to develop these regions must look at the potential utility of secondary vegetation. This chapter looks at the use of this type of forest by the Wayapi Indians of French Guiana and Brazil.

Scholars have pointed out that forest people rely to a significant degree on small game that is concentrated in disturbed forest (Vickers, 1980). Other authors (Linares, 1976) have even shown that some populations systematically adjust their hunting strategies toward gardens (in other words, slash and burn gardens and/or secondary forest). Moreover, Balee (1987) has demonstrated that garden lands and disturbed forest areas act as reserves that are not depleted due to the periodic hunting in the distant primary forest.

Few available works combine a faunal and floristic approach to the study of the utility of the secondary forest. Anderson and Posey (1985) have demonstrated that the management by the Kayapo Indians of forest plots in the savannahs have two aims: 1) to maximize the botanical diversity of useful plants; and 2) to attract game.

This chapter will present a different situation demonstrating that, with practically no mangement, the secondary forest plays an important role in the optimal foraging strategy of the Wayapi Indians.

ETHNOGRAPHIC BACKGROUND

The Wayapi Indians are Tupi-Guarani speakers divided into three groups: the first two are located in French Guiana, on the middle and the upper Oyapock River; the third is in the watersheds between the Amapari and the Jari rivers (Amapa, Brazil). The present estimate of the Wayapi population is 835 people.

Fleeing the Portuguese, the Wayapi migrated from the lower Xingu region of southern Amazonia to the Amapa region and to French Guiana in the eighteenth century. During the first three decades of the nineteenth century, there was a catastrophic decline in the population because of introduced diseases. To reduce contact with the outside world, the main groups remained isolated until the late 1940s. Today, the southern group lives under the control of the National Foundation for Indigenous People (FUNAI) on a reservation of 543,000 hectares, granted by the Brazilian federal government.

The French Wayapi are full French citizens, living on national lands and protected by a regional decree. Federal control prohibits any penetration of their territory by outsiders. The upper Oyapock River group, on which our work is based, grew from 197 in 1977 to 309 in 1990.

Geologically, the region is considered part of the Guyana shield. The average elevation is 150 meters and has an annual rainfall of between 2,200 and 2,500 millimeters, most of which occurs from mid-December to early July. The territory is composed of *terra firme* forest with some bare inselbergs. The major river is the Oyapock, which has numerous affluents broken by numerous rapids.

From an ecological perspective, the situation of the high Oyapock river group is the most interesting because of the lack of encroachments by gold prospectors and professional hunters. There, the Indians are surrounded by forests free from human occupation.[1]

METHODOLOGY

Field research for this paper is based on data collected in three Wayapi settlements between 1971 and 1982, combined with other data gathered during more recent fieldwork (1985, 1989, and 1990). The quantitative data about hunting and fishing yields were collected daily between April 1976 and May 1977 in the Zidock settlement, whose population was 125 at that time. During a year cycle (365 days), all game animals bagged and all fish taken by males above the age of 15 were identified and noted. Only selected samples of the catch (young and mature individuals) were weighed to obtain an average weight for all species. A parallel ethnozoological study was conducted with the help of the Paris National Museum. A complete map with a good set of indigenous typonyms had been established to locate all capture sites in terms of distance, and every evening, foragers were asked where they caught each animal. In this chapter, I will focus only on a part of this data, comparing principally the disturbed areas' production (including gardens and secondary forests) with the undisturbed primary forest's.

The research on the secondary forest's botanical importance is a cross-comparison between my personal ethnobotanical qualitative data[2] on the uses and environment of the species collected year by year among the Wayapi Indians, with the secondary vegetation forest inventories conducted by phytoecologists, J. P. Lescure (1986) and M. F. Prevost (unpublished). Plots' ages and areas are summarized below:

Plot's age (years)	Area	Scholar/date
2.5	100 m²	Lescure, March 1975
3.5	100 m²	Lescure, March 1975
4	150 m²	Lescure, October 1975
6	1000 m²	Prevost, August 1980
10	200 m²	Lescure, October 1975
23	200 m²	Lescure, October 1975
33	900 m²	Lescure, March 1975

DEFINITIONS

In the present chapter, we are concerned with two approaches.

1. With the floristic approach, I am considering forest succession after the first year of garden utilization until the forest is high enough (generally around 35 years old) to be considered "true forest" (ka'a e'e) by the Wayapi. All the successional phases are known by the Wayapi as "koke" (ancient garden).

2. In the faunistic approach, I do not rely so heavily on indigenous representation. In a united area, forest (ka'a) has been opposed to productive gardens (ko), successional phases of secondary forest (koke) and the highly disturbed forest areas entangled between the first ones. In fact, these areas are no longer true primary forest because of human pressure (cutting down fruit trees, trees for canoes, manioc beer troughs, small trees for hen shelters, or barbacots, and so on). Disturbed forest is full of gaps whose succession is similar to that of secondary forest.

This chapter uses categories that have been adapted from Prance et al. (1987) and from Balee (1986, 1987) as follows:

a = edible plants
    a1: major
    a2: minor
    a3: insignificant
b = construction material (houses, shelters, wharf, canoes)
c = technology (lashing material, glue, pottery temper, dye)
d = remedy, poisons, hallucinogens, stimulants
e = game animal food
f = others (magic, ritual, perfume, toys, fuel, and so on)

The environment divisions were defined as follows:

F.I = primary forest
F.II = secondary vegetation
    F.II^1: young swidden, 1–3 years old
    F.II^2: old swidden, 3–7 years old
    F.II^3: young fallow, 7–20 years old
    F.II^4: old fallow, above 20 years old

We note that the presence of a species in a type of vegetation only if it is usable at this stage of growth.

Finally, the following morphological categories were considered:

e = epiphyte
h = herbaceous
Tr = treelet
p = palm tree
l = liana and climber
t = tree

ETHNOBOTANY OF THE SECONDARY FOREST AMONG THE WAYAPI INDIANS

In 1980, for the three communities studied, gardens under production and secondary forest were only 2 km², divided in 1.5 km² around the present communities and .5 km² around recently abandoned communities. Basically only .25 percent of the territory foraged by the Wayapi was secondary forest.

Nevertheless in this small area, Wayapi were (and still are) able to find numerous useful species, as shown on Table 5-1.

Comparing the second line to the last line of Table 5-1, we can see

TABLE 5-1   *Wayapi Useful Species in Different Types of Environment*

	Epiphytes	Herbaceous	Trees	Palms	Lianas Climbers	Treelets	Total
Cultivated	0	17(10)[1]	7(3)	1(1)	5(3)	4(2)	34
Secondary forest[2]	0	26	111	2	33	20	192
Secondary forest and anthropized areas	0	22	35	0	15	8	80
Common to primary forest and secondary forest	0	4	76	2	18	12	112
Primary forest	20	51	311	27	82	40	531
All environments	20	90	353	28	102	52	645

[1]Numbers between parentheses indicate the number of plants able to survive in the two first stages of succession, F.II[1] and F. II[2].
[2]Surviving cultivated plants are discounted.

that 28.8 percent of the herbaceous, 31.4 percent of the trees, 32.3 percent of the lianas and climbers, and 38.4 percent of the treelets used by the Wayapi can be found in secondary forest; roughly 32 percent of the useful wild plants also occur there. But if we look at the third line, where *purely* secondary species are indicated, we find only 12.4 percent of useful species.

The difference between the two figures is emphasized by the fourth line, where the species common to secondary and primary forest are indicated.[3]

Clearly many species, in fact the majority of the species growing in the secondary forest, are also found in the primary forest. The comparison between secondary forest plot inventories and primary forest plot inventories done by the botanists (Lescure, 1986, Prevost, pers.comm.) shows that ll of these species have higher densities in the first environment than in the second; we can predict that many useful species will be collected from secondary forest. And in fact, they are: high densities and short distances are very decisive factors for interest in those species. But these figures permit us to evaluate both the occurrence of useful species in plots and their utility.

Table 5-2, based on my colleagues' results, provides a vivid but fragmentary view of the presence of useful species at different stages of forest regrowth. Methods of counting species introduce inconsistency if we compare plots four years old and ten years old, based on DBH, to plots six years old, based on diameter. But all in all, we can see the high

TABLE 5-2   *Occurrence of Useful Species at Different Stages of Regrowth*[1]

Plot Age (years)	Plot Area (sq. m.)	Number of Species	Number of Species Known by the Wayapi	Number of Species Used by the Wayapi
2.5[2]	100	27	22	11/40.7%
3.5[2]	100	16	16	13/80.2%
4[2]	150	20	17	16/80%
6[3]	1000	45	45	30/66%
10[4]	200	24	24	17/70.8%
23[4]	200	24	24	22/91.6%
33[5]	900	39	38	37/94.8%

[1]In every plot, usefulness has been considered at the specific level of growth.
[2]Individuals over 2 m high.
[3]Individuals over 5 cm diameter.
[4]Individuals over 5 m high.
[5]Individuals over 10 m high.

number of species used after the third year of regrowth, the average percentage of useful species between year 3.5 and 33, standing at 80.8 percent; thus, this figure is lower than the percentage obtained for a 100-hectare plot inventory of primary forest (93.8 percent) but with higher concentrations by species on a smaller area.

Table 5-3 gives a complementary view of all plants used by the Wayapi Indians based on their statements and my own observations. In this table, one can see that it is only after year 6 (stages 3 and 4) that secondary forest is a rich environment at specific levels. But we have to keep in mind that some species specific of stages 1 and 2 are also very important, due to the fact that they are irreplaceable.

We must also consider important use by categories. Cultivated species surviving in stages 1 and 2 are not included in this discussion. First of all, let us look at the useful species by large categories as defined at the beginning of this chapter (see Table 5-4).

THE FOOD CATEGORY

Out of the 192 species used by the Wayapi, 51 are major or minor food plants (26.5 percent). The most important species are *Lacmellea aculeata, Tetragastris altissima, Inga paraensis, I. bracteosa, I. edulis, I. alata, Pourouma*

TABLE 5-3   *Wayapi Useful Species by Stage of Regrowth and Morphological Categories*

	1 to 3 Years	3 to 7 Years	7 to 20 Years	Above 20 years
Herbaceous	16	12	9	5
Lianas	6	14	17	19
Trees	3	34	75	94
Treelets	3	11	12	13
Palms	0	1	2	2
Total	28	72	115	133

TABLE 5-4   *Number of Species by Specific Use Category*

Uses	Code	Number of Species by Use
Edible plants	a	15
major	a1	14
minor	a2	19
insignificant	a3	18
Construction materials	b	22
Technology	c	40
Remedies and poisons	d	59
Game animal foods	e	126
Others	f	39

bicolor, P. mollis, P. tomentosa, Perebea guianensis, Eugenia patrisii, and *Physalis pubescens.*

It is noteworthy that major sources of edible wild plants such as Sapotaceae and Palmae are not present in secondary forest: but *Inga* and *Pourouma* species rank at the fourth and fifth places of the wild fruits gathered by the Wayapi.

CONSTRUCTION MATERIALS

Secondary forest is an important source of construction material (22 out of 192 species, in other words, 11.4 percent) due to the fact that Wayapi cut rot-resistant trees between 10 and 20 cm—a relatively young stage of growth. Thus, this is not true for canoe wood, which is cut from adult trunks. The main species used for shelters and house construction, especially rafters and beams are *Tapirira guianensis; Guatteria chrysopetala; Trattinickia demerarae; Tachigali paniculata; Casearia javitensis; Qualea coerulea; Xylopia longifolia; Guatteria discolor; Tetragastris* spp.; *Sapium ciliatum; Eugenia patrisii.*

Except for *Eugenia patrisii,* hardwoods suitable for house posts are found in primary forest.

TECHNOLOGY

A myriad of uses are covered by the term technology. Forty out of 192 species are found in that category (20.8 percent).

The secondary forest provides material for dyeing not found in primary forest including *Miconia punctata, Myriaspora decipiens, Henriettea succosa, Licania heteromorpha,* and *Inga* spp.

Technology includes other uses, ranging from arrow points (*Eugenia patrisii, Calyptranthes amshoffae, Myrcia saxatilis*) to polish (*Pourouma guianenis*) and lashing material (*Trema micrantha,* various Annonaceae, *Merrimia macrocalyx,*various Lecythidaceae), materials for making spoons (*Ambelania acida, Hirtella bicornis*), and even basketry material, furnished by *Ischnosiphon obliquus* and *I. arouma,* which are more commonly found in undisturbed areas.

MEDICINES

With 59 out of 192, medicines account for 30.7 percent of all useful species registered for the secondary forest (Grenand et al., 1987). With the species found near the village perimeters, 52 percent of the medicinal species occur in areas directly affected by the Indians. Important remedies found there include *Mansoa* alliacea, *Guatteria discolor, Vismia cayennensis, Killinga pumila, Potalia amara, Bellucia grossulariodes, Inga alba, I. alata, Vataireopsis surinamensis, Piper oblongifolium, Sabicea glabrescens, Fagara* spp., *Petrea kauhotiana, Justicia pectoralis, Aristolochia* spp., *Gurania huberi, Omphalea diandra, Stigmaphyllon splendens, Abuta sandwithiana, Siparuna guianensis, Eugenia* sp., *Crotalaria retusa, Securidaca pariculata, Geophila tenuis, Solanum crinitum,* and *Costus* spp.

Therefore, half of these plants are important for the Wayapi.

GAME ANIMAL FOOD

This subject will be discussed in the second part of this chapter. At the moment, we have to keep in mind that 126 out of 192 species (in other words, 65 percent) are eaten by game animals.

Other uses: Species in this category account for 20 percent of species used (39 out of 192 species). Many are used as sources of fuel, including *Cecropia obtusa, Inga edulis,* and *I. bracteosa.*

Other uses are perfumes or incenses (*Burseraceae, Pharus* spp.); depilatory; bark mask (*Couratari guianensis*); fishing baits (*Phytolacca rivinoides* and *Cissus erosa*); or magic (*Caladium bicolor, Mansoa standleyi,* and *Xyphidium caeruleum*).

In using the Prance et al.(1987) method, counting major uses as 1.0 and minor uses as .5, Table 5-5 shows the most important species for the Wayapi, with values of 2.5 and more. This ranking reflects multiple uses of the prominent species.

TABLE 5-5   *Most Important Species for the Wayapi*

Eugenia patrisii	3.5
Hirtella bicornis	3.5
Inga alata	3.5
Cecropia obtusa	3.0
Inga alba	3.0
Inga edulis	3.0
I. paraensis	3.0
Tetragastris altissima	3.0
Protium glabrescens	3.0
Trattinicki demerarae	3.0
Ambelania acida	2.5
Bellucia grossulariodes	2.5
Cupania spp.	2.5
Fagara spp.	2.5
Hyeronima laxiflora	2.5
Inga bracteosa	2.5
Lecythis corrugata	2.5
Pourouma guianensis	2.5
Phytolacca rivinoides	2.5
Tapirira guianensis	2.5
Tetragastris hostmannii	2.5

HUNTING STRATEGIES AND PLACE OF SECONDARY FOREST AMONG THE WAYAPI

Clearly the Wayapi have a well organized schedule for using secondary forest. The data compare primary forest yields and those obtained in disturbed areas. In the first case, foraging activities are conducted on an estimated territory of 740 square miles; in the second case, on only 30 square miles.

From Figure 5-1 one can see a general trend in the Wayapi's foraging strategies with a high reliance from December to June on mammals (with a noticeable exception in March); a peak for birds in July (particularly, hunting of parrots and toucans); and, finally, three months (September to November) when Wayapi diet relies heavily on reptiles (iguanas and caimans) and fish.

To avoid game depletion, the Wayapi use different strategies that are not mutually exclusive. Their strategy optimizes hunting success by concentrating hunting pressure at the time each species is available. This availability of each species is linked with food and reproduction habits.

The optimization is achieved by the Wayapi exploring all the ecotones encountered in their territory with noticeable exception of inselbergs, which are said to house bad spirits and monsters.

Important too is their policy of big hunting expeditions. They have established camp sites in every corner of their foraged area, most of them accessible by canoe, from which radiate out 5- to 10-kilometer

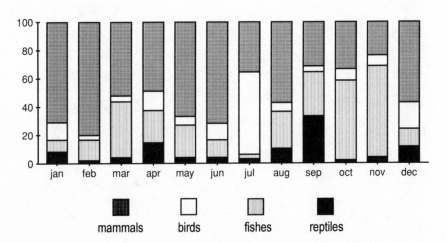

Figure 5-1 Monthly Repartition of Animal Proteins Sources (percent).

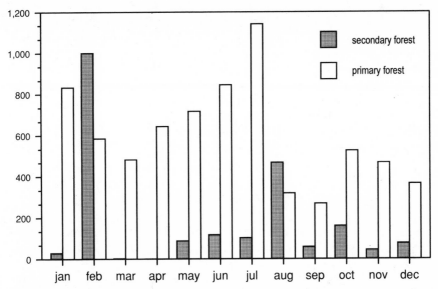

FIGURE 5-2 MONTHLY GAME PRODUCTION (MAMMALS AND BIRDS) IN PRIMARY AND SECONDARY FORESTS.

paths. This strategy differs from the trekking strategy observed by Smole (1976), Werner (1983), and the few days' travel strategy from the main villages depicted by Hames and Vickers (1982).

What then is the value of secondary forest? In Figure 5-2, we compare the monthly total production of birds and mammals (not counting fish and reptiles). It appears to have low productivity, except for two months and complementary level for six other months. In term of species, diversity is very high: we have noted 15 species of mammals (the same number as in the primary forest) and 70 species of birds (among them, a lot of small birds that are hard to see in primary forest).

What about the two most high yielding months?

In February, the high yield rate results from a collective hunt of white-lipped peccaries; this figure is random and could appear in any other month during another year. However, such a hunt is predictable every year, because peccaries are attracted by manioc and sweet potatoes surviving in the forest.

In August, the situation is quite different. The male population starts new gardens and can spend only a small amount of time hunting and fishing. Garden and secondary forest hunting appears to be the only response that fits in the work schedule. The good yields and the diversity of the catches during this month, ranging from white-lipped peccaries and red brocket deer to agoutis (mammals) and toucans, doves, and parrots (birds), show that secondary forest is considered game reserve during the rest of the year. Furthermore, the yields are increased by the trees, lianas, and treelets fruiting only in secondary forest at this time.

This is clearly an optimal foraging strategy because in September the same pattern of fruiting which would increase hunting yields in secondary forest is disregarded by the Wayapi, who turn to a new foraging strategy (that is, timbo fishing).

Eventually, to evaluate the role of secondary forest, we have to weigh up the total production of the two environments, comparing it with the areas foraged.

One view is provided by the pie charts of Figure 5-3. We can see that 27.5 percent of the game (only birds and mammals are counted) bagged by the Wayapi come from 3.8 percent of their hunting areas. If we include the boys' production (males between 9 and 14 years of age) which have not been computed here, without a doubt we can assert that over a third of the birds and mammals caught come from the secondary forest with little effort during short hunting trips.

Finally, I assume that without the high pressure that exists during August, very good yields are obtained in secondary forest.

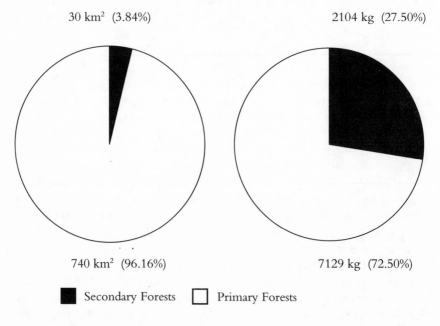

30 km² (3.84%)   2104 kg (27.50%)

740 km² (96.16%)   7129 kg (72.50%)

■ Secondary Forests   □ Primary Forests

FIGURE 5-3 COMPARED AREAS (LEFT) AND GAME PRODUCTION (MAMMALS AND BIRDS) (RIGHT) IN PRIMARY AND SECONDARY FORESTS.

CONCLUSION

What is the place of secondary forest? We know this type of forest is increasing every year in all forested regions of Central and South America. I hope that the data presented here answer this question and prove that secondary forest is not a poor environment.

Briefly, I will sum up the main points that have been made. Except for the first stage of regrowth, secondary forest contains a majority of useful plants. This is a very important fact if some are irreplaceable, useful species that cannot be found in primary-forest. We can also stress that the concentration and easy access of some primary forest species in secondary forest alleviate the human encroachment on primary forest.

The term "useful species" does not have the same meaning for all cultures; however, it was pointed out by Prance et al. (1987) that the species used by indigenous people "are far more diverse than the uses to which these species are put in western society."

Another very important aspect of the secondary forest is its place within the adaptive strategies of indigenous peoples of neotropical forests. More quantitative and qualitative data still need to be collected, since, as Balee (1985) pointed out, "There are not one, but many forests in lowland South America." This is also true for secondary forest. Our results show that secondary forest plays an important role in an optimal foraging system. It is yet another example of the in-depth indigenous knowledge about fauna and flora, demonstrating hunting practices that are clearly opportunistic, yet at the same time concerned with game conservation.

Unfortunately, I have said nothing of the present situation of the Wayapi Indians nor of other Amerindian populations in the Guianas. Some recent documents (Gallois and Ricardo, 1983; collective, 1990, and so on) have depicted a very bleak future for Amazonian Brazil. For French Guiana, we have recently shown (Grenand and Grenand, 1990) that cultural disintegration is going more swiftly.

M. Plotkin (1988) said for Suriname about its indigenous knowledge and how to preserve it, "There exists an urgent need to expand ethnobotanical research." Scientists do not need to be convinced. But what about our respective governments?

Notes

1. This is true with the noticeable exception of the presence of two small groups who wish to remain free of contact with both the Wayapi and the Westerners.

2. These data are based on 1,600 vouchers deposited in various herbariums (especially Cayenne and Paris) and determined by international specialists.
3. By primary forest, I mean undisturbed rain forest composed of high hill forest (ka'a e'e) and lower wet forest (Ka'a pe).

References

Anderson, A., and D. Posey. 1985. Manejo do cerrado pelos Indios Kayapó. *Boletim do Museu Paraense Emilio Goeldi*. Botânica 2 (1):77–88.

Balee, W. 1985. Ka'apor ritual hunting. *Human ecology* 13 (4):485–510.

Balee, W. 1986. Análise preliminar de inventário florestal e a etnobotânica Ka'apor (Maranhão). *Boletim do Museu paraense Emilio Goeldi*. Botânica 2 (2):141–167.

Balee, W. 1987. A etnobotânica quantitativa dos Indios Tembé (Rio Gurupi, Pará). *Boletim do Museu Paraense Emilio Goeldi,* Botânica 3 (1):29–50.

collective. 1990. *Brésil: Indiens et développement en Amazonie*. Ethnies: Paris.

Gallois, D., and C. A. Ricardo. 1983. *Povos Indígenas no Brasil*. CEDI: São Paulo.

Grenand, P., and F. Grenand. 1990. *Les Amérindiens, des peuples pour la Guyane de demain,* coll. La Nature et l'Homme, ORSTOM, Cayenne.

Grenand, P., C. H. Moretti, and H. Jacquemin. 1987. *Pharmacopées traditionnelles en Guyane: Créoles, Palikur, Wayãpi*. coll. Mémoires. ORSTOM, Paris.

Hames, R., and W. T. Vickers. 1982. Optimal diet breadth theory as a model to explain variability in Amazonian hunting. *American ethnobiologist,* 9 (2):358–78.

Lescure, J. P. 1986. *La reconstitution du couvert végétal après agriculture sur brûlis chez les Wayãpi du haut Oyapock (Guyane Française)*. Thèse de doctorat de l'Université de Paris.

Linares, O. 1976. Garden hunting in the American Tropics. *Human ecology* 4 (4):331–49.

Plotkin, M. J. 1988. Ethnobotany and conservation in the Guianas: The Indians of Southern Suriname, in Almeda and Pringle (ed.), *Tropical rainforest: diversity and conservation*. California Academy of Sciences: San Francisco.

Prance, G. T., W. Balee, B. M. Boom, and R. L. Carneiro. 1987. Quantitative ethnobotany and the case for conservation in Amazonia. *Conservation Biology* 1 (4):296–310.

Smole, W. 1976. *The Yanomama Indians: A Cultural Geography*. University of Texas Press: Austin.

Vickers, W. T. 1980. An analysis of Amazonian hunting yields as a function of settlement age, in Hames, R. (ed.), *Working papers on South American Indians* (N° 2):7–29.

Werner, D. 1983. Why do the Mekranoti trek? In Hames R. and Vickers, W. T. (ed.) *Adaptive responses of Native Amazonians.:* 225–38. Academic Press, New York-London.

6

The Sierra Nevada de Santa Marta: A Case Study

JUAN MAYR

Fundación Pro-Sierra Nevada de Santa Marta, Colombia

The Sierra Nevada de Santa Marta is a huge massif in northern Colombia. It rises abruptly from the shores of the Caribbean Sea, and within a distance of only 42 kilometers, its snowcapped peaks reach a height of 5,775 meters, making it the highest coastal mountain in the world.

The Sierra Nevada harbors an extraordinary array of ecosystems distributed along altitudinal gradients. During the Pleistocene era, this mountain became the refuge for numerous species, resulting in a current rate of endemism somewhere between 40 and 60 percent.

Many rivers spring from this massif, flowing in every direction, their waters irrigating the low dry plains around the great mountain. This particular hydrological system, regulated by the tropical forests of the Sierra Nevada, is intimately linked to the social, political, and economic development of the north coast of Colombia.

PRECOLUMBIAN HISTORY

In the preColumbian past, this region was inhabited by a large indigenous population, estimated at 700,000 persons, that based its economic system on the management of the ecosystems. The Indians managed these ecosystems in a sustainable manner. These peoples, the Tayrona in particular, by developing a complex system of lithic engineering, successfully managed the fragile soils, forests, and water. Thus, they avoided erosion by building terraces, roads, canals, and aqueducts.

The arrival of the Spanish in 1502 and the ensuing hundred years of violent battle brought the destruction and pillage that put an end to the splendor of the Tayronas, who abandoned their great cities of stone. Their towns were then overgrown by the jungle. The survivors of the indigenous population sought out the most remote areas of the mountains in order to continue practicing their customs and traditions. They readapted to the new circumstances imposed by the reduction of their habitat, which was now confined to the middle and upper reaches of the Sierra Nevada, where they still live to the present day.

ENVIRONMENTAL DECAY

With territorial control in European hands, new economic, technological, and philosophical concepts were imposed. Livestock was brought in, and land use changed. The subsistence economy was transformed into one of capitalist production. Thus, the systems and forms of property were altered, particularly in the lower reaches of the Sierra.

Nevertheless, all attempts at settling on the lower slopes of the massif ended in failure. This situation persisted over the course of three centuries until the early 1900s, when Capuchin missions were established in the midst of the indigenous communities, which were now prohibited from using their own language, clothing, and customs.

During this period, some of the defeated from the Thousand Day War undertook the first wave of settlement on the Sierra, where they found refuge in the mountains of the massif.

In the early 1950s, due to the political violence between Colombia's two major political parties, hundreds of poor farmers fled to the inland areas of the country, particularly to the mountains in the vicinity of the Sierra Nevada. There they found a habitat similar to their original territory with fallow lands, generally inhabited by indigenous natives. They began the process of settlement by cutting down the forests to extract precious woods, planting the first staple crops, and initiating coffee production. Later, the ancient terraces of the stone cities were plundered for their wealth of gold objects, pottery, and semi-precious stones, and there the new settlers built their homes and coffee warehouses.

Thus, a couple of peaceful decades went by in which the coffee economy was developed, until one fine day, in the early 1970s, the marijuana seed arrived along with North Americans bringing sacks full of dollars. Demand increased, as did profits. As a result of these events, from throughout the entire country came a new wave of immigrants in search of quick wealth. Their arrival brought the clearing of 100,000 hectares of forest.

The marijuana boom was so intense that many farmers replanted their coffee fields with this new crop. But the marijuana also brought violence. The first armed groups were formed for the defense, management, and export of "Santa Marta Gold," as the local crop variety was known.

Finally, the government decided to intervene. Using helicopters and crop dusters, it fumigated the illicit crops with herbicides, but it did not offer any other alternatives to the local population. In the eyes of the farmers, their only relationship with the government was through the aircraft that destroyed their economy. Some began to substitute the illicit crop with coffee, but in other cases they planted coca.

Amidst this process, fueled by violence and the discontent of the farmers, armed guerrilla groups found fertile ground on which to carry out their activities. Some locals mounted armed resistance to the guerrilla presence, and so the Sierra was torn between territories under the sway of different groups, always backed by the force of arms.

CURRENT SITUATION

Today the Sierra is a land in crisis. The lack of legislation appropriate for its particular conditions has helped make the Sierra Nevada a *de facto* state within Colombia.

The "Sierra Nevada Territorial Unit" is now divided among three departments, ten municipalities, two indigenous reservations, and two national parks. Moreover, approximately 35 government agencies attempt to intervene in the region through uncoordinated policies and programs that do not encourage sustained development.

In contrast to the internal *de facto* situation in the massif, the plains surrounding the great mountain are the most fertile lands in the country. There, the prosperity of the agroindustries that produce bananas and other fruits for export continues to grow, as do the African Palm and cotton plantations as well as cattle ranches. It is on these plains that the main economic and political centers are located, in the departmental capitals and county seats.

The causes and effects from the loss of forests are beginning to affect the normal development of the traditional inhabitants, the farmers, and the more than 1.5 million residents of the cities and towns around the great mountain, which are supplied by its waters.

The aqueducts and the irrigated agroindustrial areas are supplied by the waters of the Sierra Nevada. However, due to the growing loss of forests, the riverbeds no longer contain the flow.

Thus, at times of drought the water runs out, to the dismay of the local population and the agricultural sector; whereas in the rainy season, the great streams flood the cities and towns, destroying the roadways and communications systems and flooding vast areas of fields, producing tremendous economic losses, the magnitude of which has yet to be quantified. Moreover, the guerrilla groups that have settled in the Sierra Nevada attack the infrastructure and constitute an ongoing menace to the economic sectors of the fertile lands, leading to a complex political, social, and economic situation.

INDIGENOUS HISTORY AND PERSPECTIVE

In this regional context, the indigenous communities, ever since their very first moment of contact with the Western world, have witnessed the incessant pillage and destruction of their territories, their sacred sites, and the burial grounds and customs of their ancestors.

Of the four tribes that managed to survive the conquest, one has been integrated into the general society, while three others are undergoing various degrees of acculturation. The Kogi are the best guardians of the knowledge of their ancestors.

Of the outside actors—beginning with the wars of conquest and later the Capuchin missions, up to the settlements, narcotics traffic, guerrillas, army, and white civilization of today—none have understood the value that this philosophical store of knowledge holds for humanity.

At present, the Kogi, Arhuacos, and Azarios are practitioners of the "Law of the Mother." This is a complex code of rules that regulates human behavior in harmony with the plant and animal cycles, astral movements, climatic phenomena, and the sacred geography of the massif. The strict observance of this complex code of knowledge by indigenous society has enabled the native population to survive and remain self-sufficient over the course of several centuries. However, this unique example of harmony between humans and their environment is beginning to fade under outside intervention and the loss of fertile lands.

CONCLUSIONS

Social and economic development in the departments of Cesar, Guajira, and Magdalena is dependent on the treatment of the complex social and

ecological conditions that affect the Sierra Nevada, on the support offered to the indigenous communities in a way that respects their development models and their lands, and on the establishment of equitable
economic development.

Under these conditions, forest conservation, riverbed maintenance
and regulation, and efficient water resource management all take on
great economic and social importance.

Both the formal and informal economic sectors need one another in
the case of the Sierra Nevada. Their only established link is through
their use of the water, which makes forest conservation imperative, and
which in turn requires a social accord.

As such, the water, a product whose value extends beyond the forests, is a basis for dialogue itself between the various groups in conflict.
Without social accord, conservation of the forests and sources of water
for the future development of the region will not be possible.

7

Traditional Knowledge, Conservation, and "The Rain Forest Harvest"

DARRELL A. POSEY
Brazilian National Council for Science and Technology/CNPq,
Museu Paraense Emilio Goeldi, Brazil

Conservation has always been a cultural question, although environmentalists have acted for decades as though the preservation of nature had nothing to do with the human species. Humans are seen as innately destructive and set in some cosmic world in absolute opposition to Nature.

Most experienced environmentalists and conservationists have discovered that unless people have a direct stake and interest in conservation, then the best designed projects in the world stand little chance for long-term success. Meanwhile, scientists have begun to demonstrate how native peoples can teach us new models for sustained natural resource use and management. Their ancient indigenous traditions, developed through millennia of experience, observation, and experiment, are extremely relevant in providing future options for sustainable natural resource management.

Traditionally, Indians are thought of as merely exploiters of their environments—not as conservers, manipulators, and managers of natural resources. Researchers are finding, however, that presumed "natural" ecological systems in Amazonia are, in fact, products of human manipulation. Likewise, old agricultural fallows are extensive and reflect human-engineered genetic diversity. Since aboriginal populations were many times larger than today's indigenous populations, the extent of human influence can easily be underestimated.

THE KAYAPO AND NATURAL RESOURCE USE

For nearly 15 years, the author has coordinated an interdisciplinary team of researchers to document the knowledge of the Kayapo Indians of Brazil's Para State. The Kayapo have an integrated system of beliefs and practices. In addition to the information shared generally, there is a unique knowledge held by specialists in such areas as soils, domesticated plants, different groups of animals, medicines, and rituals. But each Kayapo believes that he or she has sufficient knowledge and the ability to survive alone in the forest indefinitely. This offers great personal security and permeates the fabric of everyday life.

Over 76 percent of the plant species used by the Kayapo are not "domesticated," nor can they be considered "wild" since they have been systematically selected for desirable traits and propagated in a variety of habitats. During times of warfare, the Kayapo could abandon their agricultural plots and survive on the semidomesticated species that had for millennia been scattered in known spots throughout the forest and savannah. Old agricultural field sites became hunting preserves and orchards, since as younger fields they had been planted by the Indians to mature for such purposes. In other words, agricultural plots were designed to develop into productive agroforestry plots dominated by semidomesticated species, thereby allowing the Kayapo to shift between being agriculturalists and hunter-gatherers. Such patterns appear to have been widespread in the lowland tropics and make archaic the traditional dichotomies of wild versus domesticated species, hunter-gatherers versus agriculturalists, and agriculture versus agroforestry.

Indigenous manipulation of environments in Amazonia was significant in molding other ecological systems. In the formation of "islands of forest" (Apete) in the campo-cerrado, for example, the Kayapo were found to have concentrated plant varieties collected from an area the size of Western Europe in to a 10-hectare plot of Apete.

Thus one of the major lessons learned from research in the last decade is that apparent natural landscapes in Amazonia may be the products of vast aboriginal populations or even diversified manipulation of ecological systems by less populous historic populations. Far too many biologists and ecologists, however, ignore the archaeological, historical, anthropological, and ethnobiological literature linking biodiveristy with human activity. Credible research in the future will have to correct this situation.

Native peoples like the Kayapo know how to both utilize and conserve the forest, which is currently being destroyed, in part, for lack of knowledge about how to utilize its vast diversity of medicines, foods, natural fertilizers, pesticides, and so forth. Fundamental to the preservation of the forest and those peoples living in it is to show that standing, living forests are more valuable than cut, burned ones. The sad truth is that currently the forest is considered economically valuable for cattle, lumber, and gold—all of which are attained only through the destruction of tropical forests and savannahs. Indigenous peoples can teach us how to give greater value to the living tropical forest—but only if they survive and we can learn to give them equal status in the future of this planet.

RAIN FORESTS HARVEST—FOR WHOM?

Industry and business discovered many years ago that indigenous knowledge means money. In the earliest forms of colonialism, extractive products were the basis for colonial wealth. More recently, pharmaceutical industries have become the major exploiters of traditional medicinal knowledge for major products and profits.

The annual world market value for medicines derived from medicinal plants discovered from indigenous peoples is $43 billion (U.S. dollars). Although no comparable figures are published for natural insecticides, insect repellents, and plant genetic materials acquired from native peoples, the annual potential for such products is easily that of medicinal plants. Research into these natural products is only beginning, with projections of their market values exceeding all other food and medicinal products combined. The international seed industry alone accounts for over $15 billion per year, much of which derived original genetic materials from crop varieties that have been selected nurtured, improved, and developed by innovative Third World farmers for thousands of years.

Likewise, natural fragrances, dyes, and body and hair products are coming to account for major world markets. Figures from The Body Shop, considered to be one of the new success stories of international enterprise, show an annual sales of $90 million with a growth rate last year of 60 percent. The 300 Body Shop products are derived from plants, are not tested on animals, and mostly come from Third World countries. These products are marketed as coming from ecologically sustainable projects managed by the native peoples themselves. The success of Anita Roddick, founder of the 14-year-old British company, has earned her the title of Britain's "Retailer of the Year." Such renown will not go unnoticed by the hundreds of would-be clones that will appear to take up Roddick's marketing strategy.

Growing interest and catapulting markets in natural food, medicinal, agricultural, and body products signal increased research activities into traditional knowledge systems. Now, more than ever, the Intellectual Property Rights of native peoples must be protected and just compensation for knowledge guaranteed. We cannot simply rely on the good will of companies and institutions to do right by indigenous peoples. If something is not done now, mining of the riches of indigenous knowledge will become the latest—and ultimate—neocolonial form of exploitation of native peoples.

Ecologists too are justifiably concerned with the ecological impact of production of "natural products" that become too successful. The tendency is always toward monocultures of cash crops. Many worry that international demands may spell the end of biodiversity, rather than encourage conservation of natural resources as initially desired.

Provocations of cultural changes can be equally disconcerting. By establishing mechanisms for just compensation of native peoples, are we not also establishing mechanisms for the destruction of their societies through the subversion of materialism and consumerism?

Given current realities, such concerns are reduced to romantic notions. The fact is that indigenous societies and their natural environments are being destroyed by the dramatic expansion of industrialized society *now.* And pharmaceutical companies and natural products companies have tasted success in their efforts; they will not go away.

Certainly the mechanisms of *what* is "just compensation" and *how* such benefits would be distributed open a Pandora's Box. But to not open this box is to accept the ethical and moral responsibility of paternalism (those from "advanced societies" know what is good for the "natives" because we have already made the mistakes of squandering our cultural and natural wealth) that has undermined indigenous independence since the first wave of colonialism.

Native peoples must have the right to choose their own futures. Without economic independence, such a choice is not possible. The current devastation of native peoples and the ecological systems that they have conserved, managed, and intimately known for millennia, require that new and drastic steps be taken to reorient world priorities. All channels and organizations—governmental, nongovernmental, professional, business—must work together to reverse the current momentum in loss of cultural, ecological, and biological diversity of this planet.

Traditional knowledge certainly provides some of the missing links necessary in conserving tropical ecosystems. We are far away, however, from having the sophistication necessary to deal with the scientific value of our exotic neighbors. We are even less capable of finding ways of sharing with them the economic fruits of our societies, or even those harvested from their own indigenous knowledge systems.

Yet until local communities see the need to preserve their surrounding environment—and have economic incentives to do so—there will never be sufficient law enforcement officers in the world to protect the best-planned parks and ecological reserves. Conservationists not only need to seek new models for natural resource management from native peoples, but must also include them in all levels of planning, decision making, monitoring—and reaping the benefits from such conservation.

II

The Potential of Nontimber Forest Products

8

Tropical Botanical Extractives

JAMES A. DUKE

U.S. Department of Agriculture

Disappearing at the rate of 11 million hectares a year, the tropical forests represent different things in the eyes of different beholders. For example, some eyes see the source of Brazil nuts and rubber for export; some see the source of food, fiber, and fuel for subsistence farmers; some see an impediment to soybean or cattle farming; some see an intriguing habitat with incredible species diversity; some see a jungle teeming with human-eating animals and fish, venomous snakes, and swarming insects; some see a protector of watersheds; some see an impediment to a dam project; some see the source of valuable timbers hopelessly enmeshed with too many "useless timber species"; some see a source of natural antioxidants, colors, dyes, medicines, and pesticides; some see a pristine agroforestry scenario; some see both the sink and the source of greenhouse gases; and still others see an underpopulated wasteland, ready to convert to progressive cities (Maeglin et al., 1990). Regardless of which eyes view this forest, there is a wealth of raw materials therein, which can be rationally exploited so that they yield employment and income to tropical forest residents while preserving the forest as an aesthetic, conservational, and economic entity.

This wealth of raw materials in the rain forest spawned the intriguing concept now entitled *extractivism*. The idea of extractivism—the renewable harvesting of economically useful products from natural ecosystems, as opposed to agroecosystems—is not new (Duke, 1981). What is new is the proposal of extractivism as a potential savior of the rain forest. Statistics, some perhaps suspect, keep cropping up, suggesting that economic incentives can make the forest look much more attractive to hungry Third World citizens than the savannahs that so often follow the clearing of the forest in Latin America. The healthy rain forest is economically much more tangibly valuable than it is "on the chopping block."

FOOD

Strident challenges by conservationists and economic botanists are leading more and more people to "think green" when marketing foods. From Annato to Brazil nut ice cream to Carambola and Zapote, more and more rain forest items are showing up in specialty stores and supermarkets (UNCTAD GATT, 1990). Tropical fruit flavors are also on the upswing. In 1989 flavor compounders said that tropical fruit flavors topped the list for innovations.

The Peach Palm (*Bactris gasipaes*), a prime source of palm hearts, is a prime example of a forest species, long utilized by the Amerindians, now finding its way out of the forest and on to urban dinner tables. This tree can furnish fruits renewably year after year, and, under "ratooning," can furnish palm hearts for more than one season.

PALM OILS

Although recently there has been some strong negative press about palm oils as foods (reports have said it is high in saturated fats), there is at least one wild palm, the "milpesos" (*Jessenia bataua*) of Panama and Colombia that yields an oil that has a composition similar to olive oil (Balick and Gershoff, 1981). The oil of the "milpesos" quite possibly has the same health benefits as olive oil, yet can be produced for less than half the price. Tankersley and Wheaton (1983) reported that in 1978 the United States imported $228,900,000 of coconut oil, $74,900,000 of palm oil, $34,800,000 in palm kernel oil, and only $36,800,000 in olive oil.

FODDER

Renewable sources of fodder for edible animals and beasts of burden are of great importance. Even in marginal seasonally deciduous forests, such plants as the Latin American "ramon" (*Brosimum alicastrum*) are lopped to provide supplemental fodder to farm animals. Under some circumstances, ramon plantations could produce almost twice as much forage as established pastures (Duke, 1989). "Living fence posts" often provide supplemental shade and fodder. Of course, the forest provides food for most, if not all, of the preferred herbivorous animals hunted for food in the extractive reserve.

FIBER

Fewer and fewer primitive people make clothing of bark. Bark cloth was made from such tropical genera as *Brosimum, Castella, Cecropia, Ficus, Hibiscus, Muntingia, Poulsenia,* and *Pseudolmedia* (Duke, 1986). Fibers from these and other tropical trees were also twisted together to make rope. Subtropical species of Hibiscus provide us with kenaf (*Hibiscus cannabinus*), which is being considered as a new crop alternative fiber in the United States. Sepals of roselle, *Hibiscus sabdariffa,* provide the red color and part of the pleasant acid flavor to popular herb teas. These Hibiscus can be grown destructively as monocultural row crops or constructively in agroforestry scenarios.

FUEL

Paleolithic humans learned early to use firewood. Early humans learned the utility of candlenuts (*Chrysobalanus, Dialyanthera, Dipteryx, Elaeis, Licania,* and *Virola,* among tropical American genera). Modern humans, worried about the dietetic consequences of saturated fats in palm oils, are nonetheless looking into palm oils (and other oilseeds) as possible alternatives to petroleum. Properly managed fuelwood farms and oilseed farms could sustainably fuel the tropical Third World economies. Some optimists have suggested that two billion hectares of tropical land producing 25 barrels of palm oil per hectare, renewably, could satisfy the world's fuel needs (Duke, 1988). Babassu palm, from tropical America, and Nypa, from tropical Asia, have also been suggested as fuel sources for the decades when petroleum runs out. Nypa is estimated to give renewably up to 90 barrels of ethanol per hectare, three times as much as sugarcane. Presently less than 1 percent of America's fuel imports are derived from tropical forests, but substituting tropical oils and alcohols, developed renewably from tropical agroforestry biomass farms *could* satisfy 100 percent.

FLORICULTURE

According to the International Trade Center (UNCTAD/GATT, 1987), total world imports of cut flowers, foliage and plants amounted to $2.5 billion in 1985. Developing countries' share of the market in cut flowers rose from 19.6 percent in 1981 to 22.8 percent in 1985, cut foliage rose from 12.2 percent in 1981 to 16.4 percent in 1985, and plants rose from 5.4 percent in 1981 to 22.8 percent in 1985. As exporters, Holland led with 65 percent of the market, followed by Colombia with 12 percent, Israel with 5.7 percent, and Italy with 5 percent. Imports of tropical plants into the United States continue to rise. In 1985, Costa Rica provided more than 54 million plants, Guatemala more than 38 million, Honduras more than 12 million, Jamaica more than 8 million, Colombia more than 5 million and so on. All together the United States imported 162.8 million cuttings of tropical foliage and young plants in 1985.

ANTIOXIDANTS

Antioxidants could easily be produced botanically in an extractive reserve. Studies have shown several botanical alternatives to be fully competitive in their activity with the synthetics. The food industry consumes only 13 percent of the antioxidant demand, 20 percent going to resin, 21–27 percent to plastics, 44–51 percent to rubber, 11 percent to functional fluids, and 4 percent spread over agricultural, cosmetic, pharmaceutical and other small industries. Butylated hydroxytoluene (BHT) is still the leading phenol antioxidant, but it has discoloring, volatility, and toxicity problems. Many antioxidant spices, like turmeric, are tropical. If Duthie's Cumulative Antioxidant Index (1991) proves out as a preventive for America's major killers heart disease and cancer, we can anticipate increased demand for natural antioxidants. I suspect herbal antioxidants now account for considerably less than 10 percent, but speculate 100 percent of the demand for vegetable antioxidants could be satisfied by tropical agroforests and extractive reserves.

BEVERAGES

It is strange how the world has leaned toward caffeine-containing beverages, all of which could be produced in the tropics. The superstars are chocolate, coffee, cola, guaraná, mate, and tea.

Chocolate, guaraná, and mate are native to tropical America, coffee is native to Ethiopia, cola to Africa, and tea to the Orient. Though few of these are now harvested from the wild, all could be grown in soil-preserving agroforests, with mate and guaraná harvested renewably from extractive reserves.

CHLOROPHYLL

Chlorophyll a (from spinach) was listed at $31.80 per milligram in the 1989 catalog of the Sigma Chemical Company and chlorophyll b was listed at $34.18 per milligram. Thus a kilogram of chlorophyll is estimated to cost $32–34 million at Sigma prices. Assuming this price and unlimited markets, each hectare of tropical forest is worth at least $50 million. Annual leaf litter in a tropical forest may run about 5 metric tons dry matter per hectare per year. Assuming a rather conservative 0.1 percent chlorophyll for the leaves, 5,000 kilograms of leaves contain 5 kilograms of chlorophyll worth more than $150 million per hectare. Today, most chlorophyll is processed from alfalfa, nettle, or spinach, but 100 percent could be a byproduct of leaf-protein, processed from leaves of tropical forests.

ENZYMES

Proteolytic enzymes may have many applications, and many such enzymes, such as *artocarpin, bromelain, calotropin, chymopapain, dumbcain euphorbain, ficin, mucunain, papain, pinguinain,* and *zingerbain,* can be derived from extractive reserves. Phillips (1990) reports on the gallery fig, *Ficus insipida,* most widespread of the neotropical nonstrangling figs. There is currently a worldwide market for its dry latex that is used, like papain, as a digestive, a chill-proofing agent in beer, a meat tenderizer, and a vermifuge. As the tree grows rarer in Amazonian Peru, the crowding together of once isolated families increases the need for the anthelmintic latex. Though a renewable resource, like rubber, this latex must be tapped cautiously with conservation in mind.

Papaya leaves have also found their way into herb teas at the health food bars, again because of the digestive properties of the proteolytic enzymes.

FOOD COLORINGS

The renewable tropical dye annatto (*Bixa orellana*) is used to color cheese, bakery products, soups, sauces, pickles, smoked fish, and other foods with low or zero fat content. (The use of annatto in cosmetics is limited because it tends to fade.) Tropical turmeric is combined with oil-soluble annatto as a margarine colorant. Both are cost-effective alternatives to beta-carotene, another widely used natural colorant.

Recently, interest in natural colors has doubled. Japan has nearly half of the world's total patents on natural colorants, followed by the United States. Of the dozens of colorants patented, "only colorants derived from cochineal, annato, and saffron are approved by the United States Food and Drug Administration" (Francis, 1987). The increasing demand for natural food dyes suggests that this small market might grow. Natural food colors make up more than 95 percent of the 100,000 tons of food colors, but certified colors consume 25 percent of the projected $180 million food color market. The beverage industry is the most important market for food colorants, accounting for 55.4 percent of all colors used, with meat and meat products next largest at 10.9 percent. Currently, I estimate that 10 percent of the market for natural food colorants is satisfied by tropical forest products.

MEDICINALS

Many have heard claims that 25 percent to 50 percent of modern medicines come from tropical rain forests. Those claims are exaggerated. I compiled a list of 250 important medicinal plants of the world (Duke, 1990). Half of the species are temperate. Of the 40 tabulated as Flowering Plants as Sources of Useful Drugs in the United States (Farnsworth and Soejarto, 1985), about half are temperate. If half of important medicinal species are temperate, we should calculate that 12.5 percent of prescription drugs contain at least one ingredient derived from tropical higher plants. In many, perhaps half the cases, one or more ingredients in that same prescription derive from minerals, synthetics, or lower plants, bacteria, or animals.

Christine Padoch (1987) notes that one crude drug, sarsaparilla (*Smilax*), commanded 200 times as much in New York as it brought the Amazonian collector. Sarsaparilla is just an over-the-counter herb, not a finished pharmaceutical prescription. Finished pharmaceuticals are often several magnitudes more expensive than the crude plants from which they derive. I estimate that crude drugs would cost closer to 1/1000 than to 1/100 the cost of finished prescription drugs.

Approximately $100 billion worth of prescription pharmaceuticals go to the wealthy 10 percent of the world's inhabitants. Many of the remaining 90 percent of the population, especially tropical citizens, derive much of their medicine directly from the environment. We must assign the same value to their lives that we assign, via prescriptions, the lives of the wealthy. Wealthy and poor alike make contributions to and extractions from society. The poor 90 percent cannot afford the $100 billion end-product pharmaceuticals, even though they may have harvested and exported raw materials at a value closer to $6 million. On top of this paltry $6 million, we must add the locally used traditional medicines. Whether we value those traditional medicines at $1, $10, $100, or $1,000 per capita year, we are talking about billions, not millions of dollars. If medicines of the wealthy 10 percent cost $100 billion, and we assign equal per capita values to the medicines of the 90 percent who cannot afford prescription pharmaceuticals, we are talking about $900 billion.

SPICES

U.S. spice imports reached an all-time high in 1987, at nearly $439 million (Foreign Agricultural Service, 1990). Of the tropical spices imported to the United States, only allspice, capsicum, and vanilla are native to tropical America, but cardamom, cloves, ginger, mace, nutmegs, black pepper, and turmeric are widely grown in tropical America. Most of these are renewable resources; however, harvesting the barks of cassia and cinnamon can be destructive. With the possible exception of allspice, few or none of the tropical spices are harvested wild from extractive reserves.

SWEETENERS

Americans may consume the equivalent of 150 pounds of sugar a year, with noncaloric sweeteners on the upswing. Intense sweetener consumption in the United States is approaching the equivalent of 12 kilograms per person per year, with 75 percent going to soft drinks, 10 percent to diabetic foods, 10 percent to tabletop sweeteners, and so on (Chemical Marketing Reporter, November 21, 1983). Diet drinks were projected to account for 25 percent of the total soft drink business in

1984 and 50 percent by 1990. One company alone approached three-quarter billion dollars per year (1985) with its aspartame. Saccharin consumption is about 3,000 tons per year, with one third of it imported (Chemical Business, January 1986). Small wonder there is such interest in tropical sweeteners derived from *Dioscoreophyllum* (Africa), *Lippia* (America), *Momordica* (China), *Stevia* (America), *Synsepaum* (Africa), and *Thaumatococcus* (Africa). Although only two of these are American, all could be grown in tropical America in extractive reserves.

VITAMINS

There is a $5 billion a year market for nutritional supplements in the United States, much of it going to vitamins. Vitamin A does not occur naturally in plants, but its precursor, beta-carotene, is abundant in green leafy vegetables. How ironic then that thousands of tropical children die each year of xerophthalmia, due to vitamin A deficiencies, when the beta-carotene in some of their tropical weeds, like amaranth, poke, and purslane, could clearly prevent this. Although most vitamins could be obtained from rain forest fruits and nuts, economics usually leads to a cheaper synthetic.

GREEN CONSUMERISM

Perhaps you believe, as do I, that using living biomass instead of fossil biomass (like asphalt, coal, petroleum, and tar) for the production of fuels and petrochemicals can lessen the likelihood of global warming. Granted, if you burn a tree instead of petroleum to heat your home, you are still generating heat and carbon dioxide. But Mother Nature abhors that vacuum generated by the removal of the terrestrial fuel (the tree) and fills the vacuum with another tree or plant life, which in turn takes back some or all of the CO_2 generated in burning the firewood. In a sort of Gaian homeostasis, nature rapidly plugs the holes left by harvesting biomass. So plant-derived fuels, like biogas from animal manures, butanol from surplus potatoes, ethanol from surplus grain, firewood from tree-falls, fuel from municipal wastes, methane from city leaves, methanol from eucalyptus, and so on, can help decelerate the global warming. First conserve—then use living instead of fossil plants where energy is necessary and obese human energy cannot do the trick. Yes, I am saying use the forest—but use it wisely. The more you increase the relative value of the forest, the better our chances to convince hungry farmers in developing countries that the forest can be worth more to them "on the hoof" than "on the slab." If Brazil Nuts and rubber alone make the standing irreplaceable forest worth more than the cattle farm or soybean farm (Schwartzman & Allegretti, 1987) that might replace it, how much more valuable can green consumerism make that forest?

CONCLUSIONS

I have enumerated some important and some trivial plant products that could be produced in the tropics. All too often, synthetics have supplanted some of these natural forest products. It is possible, but not probable, to produce all foods, fodder, fibers, and fuels from tropical agroforests. As the petroleum runs out, I visualize the forest providing more and more organic natural products. Progressive planning and pilot studies today could better prepare us to meet our needs with tropical forest products tomorrow. Subsidized extractive reserves could contribute many of the products enumerated while simultaneously performing useful, aesthetic, conservational, and ecological functions.

References

Balick, M., and S. N. Gershoff. 1981. Nutritional Evaluation of the *Jessenia bataua* Palm: Source of High Quality Protein and Oil from Tropical America. *Econ. Bot.* 35:261.

Blicher-Mathiesen, U., and H. Balslev. 1990. *Attalea colenda* (Arecaceae), a Potential Lauric Oil Resource. *Econ. Bot.* 44(3):360–67.

Brook, J. 1990. Harvesting Exotic Crops to Save Brazil's Forest. N.Y. Times, Apr. 30.

Duke, J. A. 1981. Magic Mountain, 2000 A.D. Paper No. 2. 97th Congress (1st Session), "Background Papers for Innovative Biological Technologies for Lesser Developed Countries." (OTA Workshop, Nov. 24–25, 1980.)

Duke, J. A. 1986. *Isthmian Ethnobotanical Dictionary.* 3rd ed. Scientific Publishers. Jodhpur, India.

Duke, J. A. 1988. A Green World Instead of a Greenhouse. (Updated from The International Permaculture Seed Yearbook, 1985.) Earth Island Institute. 3(3):29–31.

Duke, J. A. 1989. *CRC Handbook of Nuts.* Boca Raton, FL: CRC Press.

Duke, J. A. 1990. Promising Phytomedicinals, in Janick, J. and Simon, J. E. (eds.), *Advances in New Crops: Proceedings First National New Crops Symposium.* Indianapolis, October 23–26, 1988. Portland, OR: Timber Press.

Duthie, G. G. 1991. Antioxidant Hypothesis of Cardiovascular Disease. *Trends in Food Sc. & Tech.* 2(8):205–207.

Farnsworth, N. R., and D. D. Soejarto. 1985. Potential Consequence of Plant Extinction in the United States on the Current and Future Availability of Prescription Drugs. *Econ. Bot.* 39(3):231–40.

Foreign Agricultural Service. 1990. *U.S. Spice Trade.* F.A.S. Circular Series FTEA–1–90. USDA, Washington, D.C., April 1990.

Francis, F. J. 1987. Lesser Known Food Colorants. *Food Technology* S41(4):62–68.

Maeglin, R. R., R. L. Youngs, and J. A. Duke. 1990. *Tropical Forest Use in a World of Critical Environmental Concerns: Underutilized Species.* (Draft submitted to the Journal of Economic Botany.)

Padoch, C. 1987. The Economic Importance and Marketing of Forest and Fallow Products in the Iquitos Region, in Denevon and Padoch (eds), *Swidden-Fallow Agroforestry in the Peruvian Amazon Vol. 5 Advances in Economic Botany.* New York Botanical Garden, New York.

Phillips, O. 1990. Ficus insipida (Moraceae): Ethnobotany and Ecology of an Amazonian Anthelmintic. *Econ. Bot.* 44:534–36.

Schwartzman, S., and M. H. Allegretti. 1987. *Extractive Production in the Amazon and the Rubber Tappers' Movement.* Environmental Defense Fund, Washington, D.C.

Tankersley, H. C., and E. R. Wheaton. 1983. Strategic and Essential Industrial Materials from Plants. OTA, 170–77.

UNCTAD/GATT. 1987. *Floricultural Products: A study of Major Markets, International Trade Center,* Geneva.

UNCTAD/GATT. 1990. *Tropical Fruit Juices and Pulp.* International Trade Center, Geneva.

9

Lecythidaceae and Myristicaceae: Two Tropical Families Providing Important Secondary Forest Products

CONSTANZA LA ROTTA
Universidad Nacional Autónoma de México

Secondary forest products deserve special attention because they can often be used without damaging the ecosystem. Furthermore, the importance and potential of these products have not been evaluated systematically, and many are still entirely unknown to us.

Nontimber production in Mexico can be broken down as follows: Puebla produces rhizomes, allspice, and date palm; Chiapas, palm, barbasco, and tepescohuite; Veracruz, gums, rhizomes, and other unspecified products; Oaxaca, *Chaemaedorea* palm, barbasco, and royal palm; Hidalgo produces charcoal and allspice; Jalisco, resins and other unidentified products; Michoacàn, resins, gums, waxes, rhizomes, fibers, and other unidentified products; San Luis Potosí, fibers, maguey (American agave) leaves, and palm heart; Guerrero, charcoal; Yucatàn produces oregano, damiana, jojoba, palm leaves, pepper, and huano palm (*Anuario estadístico* 1985–88, 1990).

Michoacàn has vast forest resources, ranging from forests of cold and temperate climates to hot and humid jungles. In its use of these resources Michoacàn is the third largest silviculture producer among the Mexican states, after Chihuahua and Durango. Resin accounts for almost all of the nontimber production of Michoacàn and 17 percent of its entire silviculture production. Michoacàn is responsible for 84 percent of the total general resin production in the country. The Mexican resin industry furnishes raw material for industries involved in the production of paints, insulation, rubber, pharmaceuticals, paper, food, adhesives, plastics, armaments, automobiles, and soap (Carabais et al., 1989). Gum extraction supplies two products, the raw product for natural chewing gum extracted from sapodilla, or chico-zapote (*Achras zapote*) in the humid tropical region, and the rubber extracted from *Castilla elastica, Hevea guianensis,* and *Hevea brasilensis.*

The Myristicaceae and the Lecythidaceae families are not among the nontimber products exploited either in Mexico or in many South American countries with tropical jungles.

This chapter emphasizes the importance of these two families as significant sources of secondary tropical forest products.

The Myristicaceae family has 18 genera and 300 species of tropical trees distributed in lowland rain forests; the family is exclusively tropical. The trees grow in Malaysia, New Guinea, Africa, and Madagascar, and the American tropics, particularly the Amazon River Basin.

The trees are dioecious, with alternate leaves with glands containing essences, and verticillate quasi-horizontal branches. When an incision is made in the wood, it exudes a red sap. The fruit becomes carnose as it develops, and in its maturity, it opens into two or four valves.

For quite some time all of the species were included in the single *Myristica* genus, probably because the feminine flowers and the fruit of the entire family displayed a very uniform structure. The genera confined respectively to Malaysia, Africa, and America were mainly distinguished by the differences in the structures of the staminal column, aril, and inflorescence.

Among the most common uses of the *Myristica fragans* is the production of nutmeg and mace. The pericarp of the fruit is used to make a candy, and a lard, or nutmeg butter, is extracted from the seeds to make perfume. Some species such as the *Virola surinamensis* from Brazil have waxy seeds used as a source of oils in food products and candle production. The wood is of extremely poor quality and seldom used (Heywood, 1985).

The *Myristica* genus is represented by 80 species in South East Asia, Malaysia, Polynesia, and tropical Australia. *Myristica fragans* has been grown in Guatemala in the Lake Izabal region since 1880, and in 1946 it was cultivated for commercial production in the lowland zone of Alto Verapaz.

The *Dialyanthera* genus is present in Costa Rica and Panama. There are eight *Compsoneura* species that are distributed from southeast Mexico to Brazil and Peru. Thirty-eight species of the *Virola* genus are distributed from Guatemala to Peru and southeast Brazil. Two species are in southeast Central America. The oil is used in the soap and lamp industries, and in Guatemala the wood is used for paper pulp. The *Virola guatemalenses* is found in humid forests up to 1,150 meters above sea level in Guatemala, Honduras, Costa Rica, and Panama. In Guatemala, the dry seeds are used as a flavoring in chocolate and for machine oil (Standly and Steyermark, 1946).

In addition to these uses, in the Amazon and Chocó of Colombia, this family is vital in the construction of homes and in food products and medicine. The indigenous Andoque people eat the fruits from the *Dialyanthera parvifolia* and the *Iryanthera laevis* species, after cooking them in water. The indigenous Miraña use the trunk of the same species to construct the beams and the fences of their homes; they also make ax handles from it and use the red sap to ease toothaches and apply it to wounds in order to fight skin and mouth infections (La Rotta, 1983, 1989). The Emberà Indians of the upper Baudó River eat the cooked fruit of the *Composneura atopa,* and use the latex of an undertermined species as an antidiarrheic (La Rotta, 1985).

Moreover, this family serves very important purposes in the cosmology of the indigenous Amazon peoples. These trees have guardian animals from whom permission must be sought before the trees can be collected and used. Otherwise, the society would fall victim to illnesses and attacks from the Guardians (Dueños). In order to approach these species, special rites must be performed in which offerings are made of coca (*Erythroxilum coca*), tobacco (*Nicotiana tabacum*), manioc (*Manihot esculenta*), and vegetable salt (*Matisia calimana*). This system is part of the ecological regulation practiced by the different native peoples of the tropical forest so that only those trees that need to be used get cut and collected. Waste and nonrational utilization are severly punished by the guardians. Several illnesses are attributed to the violation of the restrictions imposed by the shaman during the different times of the year that this family is collected. These illnesses are treated with the leaves from *Iryanthera ulei, Renealmia thyrsoidea,* and *Adiantum tomentosom,* and the bark of *I. ulei* is drunk in a cold infusion to fight diarrhea.

Some species of *Virola* are particularly interesting because they are the "tobacco or perfumes of the animals," that is, they have psychotropic properties and are respected by the native peoples. Only people from certain lineages who perform shamanistic functions within the society may approach these trees. Their guardians may produce psychic disorders and even death in those who are unaware of their properties and way of handling. The aromatic species of this family are also used by indigenous males as perfumes to attract their mates. This method is called "enchundular," and it causes illnesses in women and children, which is why the system is strictly enforced by the native Amazon peoples and why the misuse of these species is the object of disputes and witchcraft wars.

Different Myristicaceae species are considered to be "hot trees." Thus, they must be "cured" by the shaman when they are used in treatments for rheumatism and muscular afflictions. Table 9-1 summarizes the known uses of the Myristicaceae.

According to Schultes and Hofmann (1979), *Virola* is known as "The Semen of the Sun," and the inhalation of the dried resin is an important part of the healing ceremony in some tribes. *Virola calophylla, V. callophylloidea, V. elongata* and *V. theiodora* are the most widely used species for the preparation of the resin.

The indigenous Bora and Witoto peoples prepare a paste from the *V. theiodora* resin, and the Makú nomads of the Pirà-Paranà River in Colombia ingest the red resin directly from the bark without any preparation. They mainly use *Virola surinamensis,* and *V. peruaviana* and perhaps *V. loretensis.* There is some sketchy evidence indicating that the shamans in Venezuela smoke the bark of *V. sebifera* and dance in order to cure fevers, or that they only drink an infusion of the bark in order to find the path of the spirits.

The first detailed description and identification of the drug was published in Colombia in 1954. It was mainly taken from the Barasana, Makuna, Tukano, Kabuyaré, Kurripaco, Puinave, and other tribes of western Colombia. The drug was used in rituals for the diagnosis and treatment of illnesses, for divination, and religious purposes.

Recent studies have demonstrated that the resin is inhaled by many native groups of the Colombian Amazon, the Venezuelan Orinoco, the Brazilian Río Negro, and other areas of the western Amazon in Brazil. Its use is strictly limited by the shamans, and in some communities the drug is taken daily.

The members of the *Lecythidaceae* family are also tropical trees, the best known of which is the *Bertholletia excelsa,* which produces Brazil nuts. The family has 20 genera and 450 species and is distributed in humid tropical regions of South America, with some genera in Africa and Asia. Table 9-2 summarizes the known uses of *Lecythidaceae.*

The trees range from small to quite large, with bisexual flowers having terminal shoots. The flowers have a flashy, villous appearance due to their numerous stamens. Pollination is made possible by bats attracted to the sweet aroma of the trees. The fruits are generally thick; the outer layers pulpy, the inner ones ligneous and hard. These indehiscent fruits develop a fissure from which the seed falls. The seeds are large and lig-

TABLE 9-1   *Uses of Different* Myristicaceae *Species*

1. Medicinal Barks	*Iryanthera laevis* *Iryanthera ulei* *Virola* sp. Opopaiko (undet.) Oímeba (undet.)
2. Edible Fruits	*Compsoneura atopa* *Dialyanthera parvifolia* *Iryanthera ulei* Kopidijo (used by the Emberà) (undet.) Nigbaoue (undet.) Pirijae (undet.) Oue (undet.)
3. Household Utensils	*Iryanthera laevis*
4. Home Construction	*Iryanthera laevis* *Virola* sp.
5. Medicinal Saps	*Iryanthera laevis* *Iryanthera ulei* *Virola* sp. Oímeba (undet.) Opopaiko (undet.) Kopidijo (used by the Emberà) (undet.)
6. Magical-Medicinal Barks	*Iryanthera ulei* *Virola* sp. Guanae (undet.) Oímeba (undet.)
7. Psychotropic	*Virola* sp.

TABLE 9-2   *Uses of Different* Lecythidaceae *Species*

1. Medicinal Barks	*Lecythis usitata* Bearer of: Gavilàn (undet.), Spider (undet.), Rat (undet.) Tiger Head (undet.) Macaw (undet.) Tank Monkey (Mico Tanque) (undet.)
2. Firewood	*Eschweilera* sp. *Lecythis usitata*
3. Edible Fruits	*Cariniana* sp. *Cariniana pyriformis* Tiger Head Fruit (undet.) *Lecythis usitata*
4. Barks for Household Utensils	*Cariniana pyriformis* *Chytroma cf. valida* Bearer of: Spider (undet.) Rat (undet.)
5. Barks for Home Construction	*Chytroma cf. valida* Bearer of: Sparrow Hawk (undet.) Spider (undet.) Rat (undet.)
6. Dye	Bearer of: Sparrow Hawk (undet.)
7. Trap Construction	Bearer of Rat (undet.)

neous without an endosperm. The South American genera are *Bertholletia, Couroupita, Grias, Gustavia,* and *Lecythis.*

The most common recorded uses in the literature are the edible seeds of the Brazil nut (*Bertholletia exelsa*) and the paradise nuts (*Lecythis zabucajo*) (Heywood, 1985). The common genus in the northern coastal regions of Guatemala is *Grias,* which has two species, *G. integrifolia* and *G. gentlei.* The former is known as Jaguillo in Honduras and Morro Cimarrón in Mexico (Standley, Williams, 1961).

The *Lecythidaceae* has four subfamilies: *Planchonioideae, Foetidioideae, Napoleonaeoideae,* and *Lecythidoideae.* This family is pantropical in the fullest sense; *Planchonioideae* members are found in Africa and even Australia, and some *Foetidioideae* members range from Madagascar to Malaysia. The *Napoleonaeoideae* includes a neotropical species, *Asteranthos brasilensis,* and two genera, *Napoleonaca* and *Crateranthus* in West Africa. The four *Lecythidaceae* subfamilies are confined to the New World from the Atlantic to the Pacific Coasts. Only a few species are distributed outside the tropics, such as *Cariniana estrellensis,* which has been collected south of the Tropic of Capricorn (Prance and Mori, 1979).

Brazil nuts are of economic importance in the Brazilian Amazon, Peru, and Bolivia. Until now, they were the second largest export after rubber. In the Brazilian Amazon, 20 percent of the department of Madre de Dios is covered by forests rich in *Bertholletia excelsa,* and these forests furnish nuts for export to the United States, Great Britain, and Germany, where the seeds are consumed fresh, toasted, or as an ingredient in other foods. The oil extracted from the seeds is edible (this is known as the primary extraction), and the second extraction is used for soaps and for lamps. The residue remaining after the extraction is used for animal feed.

The other equally nutritious and tasty nut is *Lecythis usitata.* However, it is of lesser economic importance than the Brazil nut because its seeds are removed by animals before it falls from the tree and the thinness of its sack does not protect it effectively from fungus and insects. The seeds of the *L. usitata* are eaten fresh, toasted, or as an ingredient in other foods. Its oil is also extracted.

Two closely related *Lecythis* species are *L. ollaria* and *L. minor.* However, under certain circumstances the seeds are toxic due to the presence of selenium.

The mesocarp of many *Gustavia* and *Grias* species is edible. The yellow pulp around the seed of the *Gustavia speciosa* ssp. *speciosa* is eaten fresh or cooked with meat or rice.

The trunks of *Lecythidaceae* are used locally in general construction, for hand tools, and boats. *Cariniana pyriformis,* which grows in extensive areas of northeastern Colombia, has been exported to the United States and Europe, but without success due to its high silica content. Nevertheless, Ter Well (1976) indicates that silica makes marine drills resistant. The resistance of the wood in some species of *Lecythidaceae* especially *Lecythis usitata, Eschweilera odora,* and *E. jarana,* make them excellent materials for bridge construction, as are others for paper pulp.

The use of the roots, wood, leaves, fruits, and seed of Lecythidaceae has been mentioned by many authors in the field of traditional medicine. Nevertheless, the pharmacological properties of the Lecythidaceae have hardly been studied (Prance, and Mori, 1979).

In the Colombian Amazon, in the zone inhabited by the indigenous Andoque, Miraña, Witoto, Yukuna, Tanimuka, Ufaína, and Makuna, this family is often used in the manufacture of baskets, containers, and hunting traps, for building homes, firewood, and medicinal uses.

The indigenous Miraña consider the Lecythidaceae as the "bearers of the animals," and these species also have their own guardians. As with Myristicaceae, special rights are required for the collection and use of this family.

CONCLUSIONS

The exploitation of secondary forest products is limited, in Mexico at least, to temperate zones and principally to species of pine trees. Marketing of the tropical secondary species is practically nonexistent.

Production of nontimber forest products from the different Mexican states is extremely irregular. For example, in 1987 Chiapas produced 79 metric tons of Tepescohuite, but for 1988, there is no information on this production. Simarlarly, in 1986, 218 metric tons of allspice were produced but in 1987 only 55 metric tons were produced.

In order to set guidelines for the sustainable management of nontimber forest products, it is necessary to determine the increase or decrease in the production of these products.

Moreover, the production of a single species of secondary forest products in each one of these states is a constraint because inevitably it leads to the accelerated decline in the few species that are exploited. This is why it is important to diversify production and undertake exploration for other as yet unknown tropical resources.

The irregularity in the registered production statistics for secondary forest products is a serious barrier to the analysis and formulation of sustainable rational management. Yucatán is the state in which nontimber forest production is most diversified. However, the quantification of real production for each of the resources there is virtually nothing. We do not know the production level of jojoba, damiana, oregano, or other products.

Chemical and pharmaceutical research is required in the cases of Lecythidaceae and Myristicacea in order to predict resource exploitation with certainty. Furthermore, it is necessary to study the physiology and the ecology of the germination of the seeds of these two tropical families, as well as the dynamic of the populations. The absence of this information is a barrier in devising better plans for sustainable management of these important tropical trees.

References

Anuario Estadístico del Estado de Chiapas. 1990. INEGI. Gobierno del Estado de Chiapas. Mexico.

Anuario Estadístico del Estado de Guerrero. 1990. INEGI. Gobierno del Estado de Guerrero. Mexico.

Anuario Estadístico del Estado de Hidalgo. 1987. INEGI. Gobierno del Estado de Hidalgo. Mexico.

Anuario Estadístico del Estado de Jalisco. 1986. INEGI. Gobierno del Estado de Jalisco. Mexico.

Anuario Estadístico del Estado de Michoacán. 1984. INEGI. Gobierno del Estado de Michoacán. Mexico.

Anuario Estadístico del Estado de Oaxaca. 1988. INEGI. Gobierno del Estado de Oaxaca. Mexico.

Anuario Estadístico del Estado de Puebla. 1990. INEGI. Gobierno del Estado de Puebla. Mexico.

Anuario Estadístico del Estado de San Luis Potosí. 1988. INEGI. Gobierno del Estado de San Luis Potosí. Mexico.

Anuario Estadístico del Estado de Veracruz. 1984. INEGI. Gobierno del Estado de Veracruz. Mexico.

Anuario Estadístico del Estado de Yucatán. 1986. INEGI. Gobierno del Estado de Yucatán. Mexico.

Carabias, J. V. M. Toledo, C. Toledo, and C. González-Pacheco. 1989. *La Producción Rural en México: Alternativas Ecológicas.* Fundación Universo Veintiuno. Mexico.

Heywood, V. H. 1985. *Las Plantas con Flores.* Editorial Reverté. Mexico.

La Rotta, C. 1983. *Observaciones Etnobotánicas sobre Algunas de las Especies Utilizadas por la Comunidad Indígena Andoque.* Editorial Ducal. Bogota, Colombia.

La Rotta, C. 1985. *Estudio Etnobotánico de las Especies Utilizadas por la Comunidad Indígena Emerá del Alto río Baudó.* Mecanografiado. Fundación Segunda Expedición Botánica. Bogota, Colombia.

La Rotta, C., P. Miraña, M. Miraña, B. Miraña, M. Miraña, and N. Yukuna. 1989. *Especies Utilizadas por la Comunidad Miraña.* Estudio Etnobotánico. Editorial Presencia. Bogota, Colombia.

Prance, G. T., and S. Mori. 1979. *Lecythidaceae Part I (Asteranthos, Gustavia, Grias, Allantoma & Cariniana).* Flora Neotropica. Monografía 21. The New York Botanical Garden. USA.

Schultes, R. E., and A. Hofmann. 1979. *Plants of the Gods.* New York: McGraw Hill.

Standley, P., and P. J. Steyermark. 1946. Flora of Guatemala. *Fieldiana: Botany* (24) 4:294–99.

Standley, P., and P. L. Williams. 1961. Flora of Guatemala. *Fieldiana: Botany* (24) 7:1, 261–63.

Ter Well, B. H. 1976. Silica Grains in Woody Plants of the Neotropics. Especially Surinam. *Leiden Bot. Ser.* 3:107–42.

10

The History of Nontimber Forest Products from the Guianas

ROBERT A. DEFILIPPS
Department of Botany, Smithsonian Institution

Tropical countries today are fortunate to have nontimber forest products available as materials to facilitate the search for diversification of national economies. But such was not the case in Europe where a succession of civilizations collapsed due to deforestation and the associated problems of erosion, soil nutrient depletion, siltation, and flooding. The demand for timber and thus forest-cutting was heavily implicated in the demise of the civilizations of Mesopotamia (the Sumerian civilization collapsed around 2000 B.C.), Bronze Age Crete and Knossos (the Minoan civilization, *c.* 1050 B.C.), Mycenean Greek civilization (*c.* 1200 B.C.), Cyprus (abandoned *c.* 1050 B.C.), Greece and the Golden Age of Athens (*c.* 400 B.C.), the Roman Republic (*c.* 400 B.C.), Cairo (*c.* 1000 B.C.), and the Venetian Republic (*c.* 1500 A.D.).

In the New World also, trees were "judged for their service to human needs" (Perlin, 1989). In the West Indies and Brazil, wood was used in prodigious quantities to build and fuel the sugar mills. It is extrapolated that about 90 acres of forest land was stripped annually in order to provide fuel for just one mill. Hispaniola and Barbados were largely cutover, and such clearance caused lowering water tables, desiccation of rivers and streams, as well as land slippage. When the supply of local wood was exhausted, the sugar mills imported wood from the eastern forests of the United States. Later, between the years 1810 and 1867, in the eastern half of the United States, 200,000 square miles of woodland were cut for fuel, and 25,000 square miles of forest were cut for building materials. In the period 1850-1860, farmers in the United States deforested 32,250 square miles of land to plant crops.

Those figures indicate the magnitude of past demands for timber, for which a great hunger still exists all over the world. Luckily, this need is being counterbalanced by a growing effort in the developing world to explore and judiciously exploit the nontimber forest resources, particularly those of emerging nations, in order to help conserve the forests in their totality and forestall extermination of their ecosystems.

Exactly which plants are involved in the subject of tropical nontimber forest products? To help answer this question, I turned to the three-volume work on *Minor Products of Philippine Forests* (1920–1921) edited by William H. Brown. In it, Brown devotes separately authored chapters to the following subjects and species: 1) Mangrove Swamps, 2) Palms and Palm Products, 3) Bamboos, 4) Fiber Plants, 5) Paper Pulp Sources, 6) Resins, Gums, Seed Oils, and Essential Oils, 7) Wild Food Plants, 8) Natural Dyes, 9) Ornamental Plants (mostly epiphytic orchids), 10) Soap Substitutes or Scouring Materials, 11) Official (Pharmacopoeia) Medicinal Plants, 12) Poisonous Plants (including Philippine tribal arrow poisons), 13) Miscellaneous Useful Wild Plants (including firewood, ink, lye, paper substitutes, sphagnum moss, tannins, tobacco substitutes, and tree-fern trunks), 14) Edible Fungi, and 15) Medicinal (Nonpharmacopoeia) Uses of Philippine Plants. All of these are so-called "minor," or nontimber forest products representing a great diversity and potentiality.

HISTORICAL PERSPECTIVE

The Guianas are embedded high in the green shoulder of northern South America, an area once known as the "Wild Coast." They are the only nonLatin American countries in South America, and are situated just north of the equator in a configuration with the Amazon River of Brazil to the south and the Orinoco River of Venezuela to the west. The three Guianas comprise, from west to east, the countries of Guyana (area: 83,000 square miles; capital: Georgetown), Suriname (area: 63,037 square miles; capital: Paramaribo), and French Guiana (area: 34,740 square miles; capital: Cayenne).

Perhaps the earliest physical contact between Europeans and the present-day Guianas occurred in 1500 when the Spanish navigator Vincente Yanez Pinzon, after discovering the Amazon River, sailed northwest and entered the Oyapock River, which is now the eastern boundary of French Guiana. As early as 1503 French colonists attempted to settle the island on which Cayenne is built. Within the boundaries of today's

Guianas, the land was originally occupied by Amerindians of Carib and Arawak language families; from the late 1500s onward it was almost interchangeably settled by Spanish, British, Dutch, and French traders, adventurers, agriculturists, and colonists.

Gradually the land was sorted into areas controlled exclusively by either British, Dutch, or French interests, with the former British domains becoming independent on May 26, 1966 as the Cooperative Republic of Guyana, and the former Dutch domains becoming independent on November 25, 1975 as the Republic of Surinam. French Guiana became an Overseas Department of France in 1946 and is considered an integral part of France.

Most of the some 500 species of tropical ornamental plants utilized at present in the Guianas, and some of the nearly 850 medicinal plants, were introduced in historical eras, and under circumstances which seem the distance of a universe away from the conditions that often occur today. Indeed, for a long time the New World itself was only a vague vision. The earliest European acquaintance with tropical vegetation was a result of Alexander the Great's invasion of northern India in 326 B.C., at which time the banyan tree (*Ficus bengalensis*) was first observed by the Western eyes of the conqueror's Greek forces (Stearn, 1976, 1988). Information about the wondrous banyan fig tree with its dangling aerial roots flowed back to Greece and was recorded by the classical Greek scholar Theophrastus. The impressive Indian vegetation was soon largely forgotten by Europe; indeed, the literature was later sometimes suppressed for being of pagan (non-Christian) origin.

The first travellers came to South America primarily in search of gold, spices, and new souls for the Church, because in the sixteenth century, the divine scheme of the universe was the redemption of sinners in a disobedient world. Thus the discoverers, after praying to the Madonna of the Navigators for protection, sailed to bring news of "The Redeemer" to the "misguided" peoples of America.

It required many years for the world to acknowledge the existence of South America and fit it into the already established "triple world" cosmography of Asia, Africa, and Europe, an essentially Mediterranean-oriented concept of the globe. A four-part world was gradually accepted in the sixteenth century, comprising the four great land masses of Asia (formerly one-half the world), Europe (formerly one-quarter of the world), Africa (formerly perceived to be one-quarter of the world), and America as a fourth continent often signified by the River Plate (now Argentina).

As the existence of South America was finally accepted, the pace of scientific explorations and discoveries quickened, leading to works such as *Plantae Surinamenses* (1775) by Carl Linnaeus's Swedish pupil Jacob Alm (based on Suriname collections made by Carl Gustav Dalberg in 1754–1755 under a subvention from Gustav III, King of Sweden), and the *Histoire des plantes de la Guiane Françoise* by Fusee Aublet, published also in 1775.

The Amerindians originally lived in uncontaminated ecological harmony with their forested surroundings and had a thorough knowledge of the uses of plants. In contrast, the first Europeans in the area often felt themselves imprisoned in an inpenetrable and meaningless green mass, as they eked out a living from the forest. The perceived role of the human in the forest was early studied and influenced by the famous French naturalist Count George-Louis Leclerc de Buffon (1707–1788), who produced in 44 volumes the encyclopedic *Histoire Naturelle*. Buffon was the intendant (supervisor) of the Jardin du Roi in Paris from 1739–1788. During his tenure, incidentally, the South American expedition of C.-M. de la Condamine discovered a species of rubber, *Hevea guianensis,* in French Guiana.

Buffon was deeply interested "in the changes which men had made in their natural environment, particularly the transformations which had accompanied the growth and expansion of civilization and the migration and dispersion of human beings and their domesticated plants and animals throughout the habitable parts of the earth" (Glacken, 1960).

In the days of Buffon (long before the "greenhouse effect" was understood), many believed that nature is of divine origin and must be improved and arranged by humans, who are also of divine origin. This led to several erroneous theories, including the idea that humanity must aid nature by changing it through deforestation so that more of the sun's (solar) heat could warm the earth's surface, and compensate for the heat lost due to the cooling of the earth. Thus, Buffon's studies of the physical effects of humans' intervention in the world environment led him to consider the climatic changes that occurred as a result of landclearing, agriculture, and drainage, as being beneficial.

Buffon's viewpoint, as expressed by Glacken (1960), was that, for 3,000 years, "Flowers, fruits, grains, useful species of animals have been transported, propagated, and increased without number; useless species have been eliminated. Mining has advanced. Torrents have been restrained and the rivers directed and controlled. The sea has been conquered. Land has been restored and made fertile." He believed France would be much colder than it is if its forests had not been cut.

Unfortunately, for partial proof of this theory he chose to indicate "the deforestation, scarcely a century earlier, of a district around Cay-enne (there are many references to French Guiana throughout the *His-toire Naturelle*), which caused considerable differences in air temperature, even at night, between the cold, wet, dense forest into which the sun seldom penetrated and the clearings; rains even began later and stopped earlier in them [clearings] than in the forest" (Glacken, 1960).

An article on forest conservation was published by Count Buffon in 1739, as he also believed that deforestation could be reconciled with conservation under certain locally mitigating circumstances. Essentially, he felt that "large areas inimical to man had to be cleared to make the earth habitable, but once societies were established on them, the forests were resources which had to be treated with care and foresight." At the present time, the need for environmental conservation in the Guianas is being addressed by numerous researchers, and the judicious exploitation of indigenous ornamental and medicinal plants, through careful inven-tory of species and habitats, will assist the economic development of the area.

As the Europe of earlier times became intrigued by the human inhab-itants of the New World, the Amerindians were gradually incorporated to an extent into European culture. The famous natural historian Alex-ander von Humboldt, for example, noted in the early 1800s that "When we speak in Europe of a native of Guiana, we figure to ourselves a man whose head and waist are decorated with the fine feathers of the macaw, the toucan, and the hummingbird. Our painters and sculptors have long since regarded these ornaments as the characteristic marks of an Ameri-can" (Honour, 1975).

Inevitably, New World plants were destined to play a much more important role in European life than would American tribal peoples and wildlife. Useful plants of New World origin sent to Europe from 1493 onward include maize (Indian corn), tobacco, sweet potatoes, white po-tatoes, tomatoes, vanilla, cotton, chocolate, red peppers, and pine-apples. Other plants, such as annatto, performed different functions.

The annatto plant, *Bixa orellana,* which is now sometimes grown in the Guianas as an ornamental and medicinal plant, formerly was inten-sively cultivated on plantations in French Guiana as a source of red fabric dye from the seeds. In the year 1752, for example, exports of annatto from French Guiana amounted to 260,541 pounds, which outweighed the colony's combined production that year of sugar, cotton, coffee, ca-cao, and timber. All of that produce was the result of labor from 90 French families, 125 Amerindian slaves, and 1,500 black slaves (Rodway, 1912).

SURINAME AND THE NETHERLANDS

True to the spirit of the famous Netherlands Golden Age of seaborne mercantilism, trading, and investment (Koningsberger, 1967; Schama, 1987), Suriname itself was administered as a business enterprise. In the year 1684 the society of shareholders, each owning one-third of the Surinam trade, consisted of the Dutch West India Company; the City of Amsterdam; and the private possession of the family Van Aerssen van Sommelsdijk (Heniger, 1987). A number of the Society of Surinam directors were also directors of the Dutch East India Company, and commissioners of the Amsterdam Hortus Medicus, or botanical garden.

Consequently a strong interrelationship of plant exchanges developed between Suriname, the Dutch East Indies, and the Amsterdam garden (Wijnands, 1987; Stearn, 1988). Cornelius van Sommelsdijk, the Governor of Suriname residing in Paramaribo, received from the Amsterdam garden various plants such as mulberry trees, peach trees, pomegranates, and rhubarb, to be propagated on the plantations. In return, indigenous Suriname ornamental plants soon arrived in Holland, such as *Canna indica* in 1687 and the tree-cactus *Cereus hexagonus* in 1689. The introduction of Surinamese plants to the Netherlands has been accorded extensive study by Brinkman (1980).

The wild *Caladium bicolor* (Araceae) was sent from Suriname to the Amsterdam botanical garden in 1704; at least 1,500 cultivars (cultivated varieties) of this decorative plant have subsequently been developed. In Suriname, coffee was first planted in 1711 and exported in 1721; cacao was first planted in 1700 and first exported to Holland in 1706; and cotton was first cultivated on a large scale in 1763 (Van Lier, 1972). A ship's captain introduced pineapple plants (*Ananas comosus*) from Suriname to the Amsterdam and Leiden botanical gardens in 1680, and mass plantings under glass began on Netherlands estates in 1700 (Wijnands, 1983). Pineapples are today occasionally grown for ornament in the Guianas, and a beautiful rendering of the plant is found in the 1705 book by Maria Sibylla Meriam, a famous Dutch artist who originally traveled to Suriname to study tropical insects. Her sumptuous volume is entitled *Metamorphosis Insectorum Surinamensium*. Several of the plates have been designated as "type" for well-known Surinam plant species, for example, *Hippeastrum puniceum* and *Manihot esculenta* (Anonymous, 1962; Engle, 1988; Stearn, 1982; Welebit, 1988).

Certain unique pineapple plants that had leaves with the preferable smooth (rather than spiny) leaf margins were sent from Cayenne, French Guiana to France in 1820 and grown at Versailles. This Smooth

Cayenne cultivar later became the basis of the formidable Hawaiian pineapple industry, for in 1886 Captain John Kidwell planted 1,000 crowns of Smooth Cayenne in Hawaii to establish large-scale cultivation of the fruit.

FRENCH GUIANA AND FRANCE

The first plant collector in French Guiana was Pierre Barrere (1690–1755) who, as early as the years 1722–1725, botanized along the coastal area in his capacity as "medecin-botaniste du roi" in Cayenne (Vermeulen, 1985). He was later followed by J. B. C. F. Aublet (1720–1778) who collected plants from 1762–1764 as "apothecaire-botaniste du roi" in Cayenne (Bernardi, 1976; Howard, 1983; Zarucchi, 1984; Plotkin et al., 1991). Empress Josephine of France, wife of Napoleon Bonaparte, assembled a large collection of plants from around the world in her garden at Malmaison, a chateau about seven miles west of Paris (David, 1966; Lamb, 1991; Mauguin, 1933). Plants were sent to her from French Guiana by Joseph Martin, director of a national spice plantation at La Gabrielle, 20 kilometers south of Cayenne. During the Napoleonic Wars in 1803, two British privateers captured a French ship laden with 140 tubs of living plants that had been gathered during six years of work in French Guiana by Martin.

Martin was aboard the ship, and the plants were intended for Josephine's garden and the Paris Natural History Museum. In England, however, even the intercessionary influence of Sir Joseph Banks, who wanted to send the plants forward to Josephine, was unsuccessful, and three barges dragged the plants up the Thames River, to be planted at the Royal Botanic Gardens, Kew. The plants did not survive and the unfortunate prisoner, Martin, was repatriated to France, never to see the fruits of his labor again (Stearn & Williams, 1957). He is commemorated by the plant named *Bignonia martini* (Bignoniaceae).

In the Indian Ocean, the French King's Garden on Mauritius provided Cayenne with seedlings of two precious spices, clove and nutmeg, in hopes of breaking the Dutch monopoly of the spice trade in the late 1770s (Duval, 1982; Ly-Tio-Fane, 1982). As noted by Howard (1953), "with seeds or plants, new plantations could be established in other areas of the world and new supplies of spices obtained for the markets of Europe." In 1791, French Guiana supplied nutmeg plants to the St. Vincent Botanic Garden in the West Indies, and it is believed that all the nutmegs in the economy of St. Vincent and Trinidad are progeny of those originally introduced from French Guiana (Howard, 1954).

RALEIGH'S SEARCH

One of the first British adventurers to exert influence in the Guianas was Sir Walter Raleigh, Captain of the Guard to Queen Elizabeth I. Led on by samples of marcasite of dubious origin, he went searching for the site of lost gold mines said to be in a fabled new empire, which supposedly had furnished the golden walls of the Peruvian Incas' Palace of the Sun. The mines were thought to be at the golden city of Manoa, called by the Spanish "El Dorado," set beside a certain Lake Parima (Lacey, 1973). Raleigh twice visited the Orinoco River region of the then Spanish Guiana (now Venezuela), in 1595 and 1617 (Lacey, 1973; Rowse, 1962; Waldman, 1950; Winton, 1975). On the 1617 trip he traveled under the auspices of King James I, a monarch who was 700,000 pounds sterling in debt and had been promised the gold mines in return for supporting Raleigh.

Today, it is believed (Hills, 1961) that the legendary Lake Parima sought by Raleigh is Lake Amuku, a "wet weather lake" in the northern savannahs, north of the Kanuku Mountains in Guyana. It results from annual inundation of the Rupununi savannahs by the adjacent rivers, and is thus a disappearing lake that exists only at intervals.

CURARE AND MEDICINAL PLANTS

It was formerly believed that Raleigh was the first person to mention in European literature (in 1595) the existence of curare, a paralytic poison prepared for use in hunting by the Amerindians. This distinction now seems to rest with Pieter Martyr d'Anglera (McIntyre, 1947; Thomas, 1963). Curare is of two kinds, based either on *Strychnos* (Loganiaceae) or on Menispermaceae (Krukoff and Moldenke, 1938; Schultes and Raffauf, 1990). The type known as tube-curare is prepared from several genera (for example, *Chondodendron*) of the moonseed family (Menispermaceae). Purified extracts of the dextrorotatory form of tubocurarine, a chemical from the bark and leaves, are used in modern medicine as a muscle relaxant for shock treatment of mental illness, and as an adjunct to anaesthesia in heart surgery.

Using Surinam plants, Johann C. D. von Schreber in 1783 was the first person to describe precisely which plant species enter into the composition of curare; later, Richard and Robert Schomburgk identified certain plant ingredients of the curare prepared by the Macushi Indians of Guyana. Ironically, it was not until the explorations of the Schomburgk brothers in the 1830s that the legendary city of Manoa, so diligently sought by Raleigh, was finally proved to be nonexistent.

In preparing a forthcoming publication on *Medicinal Plants of the Guianas* by Dr. R. DeFilipps, Dr. Mark Plotkin (Conservation International) and Ms. Shirley Maina (Smithsonian Institution), for Reference Publications, Inc. of Algonac, Michigan, we have observed that, in addition to the many species used in preparation of curare by different tribes, the Guianan Amerindians have found many treatments for malaria, leishmaniasis, skin parasites, venereal diseases, vermifuges, and abortifacients based on plant materials. Thus, many of the species are becoming known through ethnobotanical studies of long duration.

In this connection, one may note the comments of Dr. Elaine Elisabetsky of the Laboratory of Ethnopharmacology, Universidad Federal do Para, Brazil, which were made during the Workshop on Drug Development, Biological Diversity, and Economic Growth, held on March 13–14, 1991 at the National Institutes of Health, Bethesda, Maryland. Dr. Elisabetsky noted that industrialized countries and developing countries have different development priorities.

The developed countries are currently most interested in problems of cancer, heart disease, aging, AIDS, viruses, and genetic diseases. In contrast, the major health concerns of tropical countries are diarrhea, malaria, leprosy, tuberculosis, parasitic diseases, and various sexually transmitted diseases in addition to AIDS. Also, the developing countries include the subjects of family planning (contraceptives and abortifacients) and cosmetics (soaps and lotions) as part of the totality of their health care system, whereas industrialized countries seem mostly involved with litigation over abortion rights and in treating the various diseases of affluence.

In order to emphasize the point that drug companies in northern zones should not seek to impose a north-temperate outlook on the world in connection with drug development, one of the formal observations made by consensus of the conference participants reads: "Efforts to develop drugs from medicinal plants should address diseases and health problems seen in developing countries as well as diseases which primarily affect developed countries' populations."

BRITISH GUIANA DURING THE VICTORIAN AGE

Queen Victoria (1819–1901), who reigned from 1837–1901, gave her name to the Victorian Age, a period when unprecedented discoveries and explorations were taking place all over the British Empire. A famous plant named for her is the extremely decorative royal water lily (*Victoria amazonica*), which was rediscovered on the Berbice River in Guyana in 1837 by Robert Schomburgk, explorer for the Royal Geographical Society. He considered the plant to be "the most beautiful specimen of Flora of the Western Hemisphere." In 1848 seeds were sent to Kew from Guyana, and once successfully germinated, the seedlings were distributed to other gardens. Soon afterwards, special conservatories ("Vic-Houses") intended solely to supply the space requirements of the gigantic leaves of the royal water lily were being constructed all over Europe: at Leiden, Kew, Exeter, Sheffield, Syon House (England), Brussels, Bonn, and Berlin (Hix, 1974). Keen numismatists know that the plant appears in a shield on the 10-cent and 25-cent coins of Guyana *c*. 1977–1978, and eagle-eyed motion picture fans will see the artificial, yet realistic looking, Victoria with its upturned leaf-margin in the pool of Munchkin-land by the Yellow Brick Road in the 1939 classic, *The Wizard of Oz*. The biology of this aquatic masterpiece of the vegetable kingdom has been extensively studied in the field by Dr. G. Prance, Director of the Royal Botanic Gardens, Kew.

During the late Victorian era, Guyana became richly populated with exotic plants. Reports of the Georgetown Botanic Garden indicate that in a relatively short period of five years in the 1880s (exemplified by Jenman, 1888), ornamental plant interchanges took place involving Georgetown and the botanic gardens in Trinidad; India (Calcutta, Saharunpur, Bangalore, Ootacamund, Seebpore); England (Kew, Cambridge); Jamaica; Ceylon; Singapore; Mauriţius; Australia (Brisbane, Melbourne); Dominica; and Java.

ORNAMENTAL PLANT CONSIDERATIONS

In general, when botanists consider the origin of the vegetation and plant communities (floristic composition) of an area, they do so in terms of reference to factors such as continental drift and the direction of plant migration routes; geological factors, including orogenic elevation of mountains, and soil types; climatic and biogeographical patterns; and concepts of refugia, endemism, successional changes, evolution, and extinction. For an application of these scientific concepts to the indigenous flora of the Guianas, see de Granville (1988), Lindeman and Mori (1989) and Schnell (1965).

The background of ornamental plants as a class of study, however, is of another nature: "human nature." Ornamental plants are dispersed, in contrast to natural means, by way of gardeners with a contagious enthusiasm for growing things and distributing them; by plant collectors, hobbyists, and economic botanists; by botanic gardens whose mandated purpose is often to introduce new plants to cultivation and to horticulturally improve the wild species for enhancement of the public life; and by landscapers, land developers, and town planners. On the other hand, in the case of some people, it would seem that the insatiable urge to collect and accumulate large quantities of plants is a quite naturally ordained, internally regulated instinct that has somehow mutated into a pathological obsession.

Many ornamentals are actually multiple-purpose plants that can be employed in several ways at once, without invoking the statement by the Puritan divine and clergyman Cotton Mather (1663–1728), who entreated: "That which is not useful is vicious." In addition to being decorative, which signifies the appeal of a plant to one's sense of taste, beauty, and refinement, some ornamentals are also variously used for food (fruit, storage roots, stems, leaves), fiber, oils, as shade providers, medicines, drugs (herbal preparations), fencing material, and firewood. The coconut palm (*Cocos nucifera*) is but one good example of a multiple-use ornamental plant of the Guianas.

In the old days, some ornamentals in Georgetown, Guyana were utilized in still other ways, for fire and disease prevention. The trees lining the streets and avenues in decorative fashion were useful in preventing or delaying fire from spreading from one house to another. Unlike some trees of northern climates, certain tropical broadleaf trees do not ignite unless the heat first dries them thoroughly. As for disease prevention, the planted trees were important for draining pools of stagnant water with their roots—stagnant water that was the breeding place for mosquitoes transmitting yellow fever, malaria, and elephantiasis.

In the Guianas one still finds myriads of plants indigenous to the tropical zones of Central and South America, the West Indies, Africa, Asia, Australia, and the Pacific Islands. They are favored survivors over many introduced species that were sunk into obscurity by the dictates of fashion. Biologically, the ornamental trees may be climatically adapted to a seasonality whereby in dry seasons the leaves are dropped and flowers produced; others may appear to be continually evergreen (Bernhardt, 1987; Kingdon-Ward, 1956; Lewis, 1989). These phenomena are due to their evolutionary adaptation in a particular region, often foreign, where the climatic combination of temperature, rainfall, and humidity may be either somewhat different than, or the same as, in the Guianas. The answer (at least partially) to the level of perceived compatibility or incompatibility of introduced plants in a new region may be sought in the study of homologous habitats or environments in different regions of the world, as demonstrated by the writings of V. M. Meher-Homji (1964, 1965, 1971, 1971a; also Bharucha and Meher-Homji, 1965) and L. B. Uichanco (1969).

EXAMPLES OF FUTURE POTENTIAL

While exotic plants from around the world are quite prominent in the Guianas, there are numerous indigenous plants that seem to deserve attempts at a wider dissemination for outdoor tropical landscaping. Among these are many orchids, as well as the widespread, gigantic neotropical bird-of-paradise (*Phenakospermum guyannense*, Strelitziaceae) and the equally gigantic *Brocchinia micrantha*, an arborescent bromeliad from the Kaieteur Falls area of Guyana. The horticultural potential of the Mount Roraima area of Guyana (adjacent to Venezuela) has been discussed by Alan Toogood (1983) in an article entitled *Plant Trek to the Lost World;* populations of the wild bromeliad *Vriesea splendens* growing there seem to have more visual appeal in their bronze leaf-bands than do selected forms now in cultivation.

Much effort has recently been devoted by people working in nurseries in the Guianas to the selection, hybridization, and commercial production of local *Heliconia* (Heliconiaceae) species such as *H. psittacorum* and its numerous cultivars, especially in Guyana and Suriname. Much of this work has been reported in the *Bulletin of the Heliconia Society International*, representing an organization devoted to all members of the order Zingiberales, which includes the heliconia, banana, ginger, costus, canna, maranta, and bird-of-paradise families. An all-color illustrated guide to *Heliconia* has recently been produced by Fred Berry and John Kress (1991).

At my own deliberate speed I am attempting to assist the development of a future expanded trade in ornamental plants for Suriname. I recently introduced to that country some germ plasm of the red bird-of-paradise (*Strelitzia reginae*), which will be propagated on a small parcel of disturbed savannah I acquired at Waterland, District Para. We are also growing in pots the very common, indigenous, *Phenakospermum guyannense* for eventual distribution elsewhere, as well as many small specimens of the abundant burití palm, *Mauritia flexuosa*, which deserves a wider appeal in tropical landscape plantings.

One may ask: "Why bother to propagate such exceedingly common plants?" The answer involves the fact that, aside from their intrinsic beauty and interest, they are not much known outside of South American localities, and may be good material for export to specialist hobbyists. As for the burití, it may in the future become locally rare. Evidence of this is being studied now in Guyana by personnel of an organization known as PLENTY. It has been seen that in the Arawak Indian missions of St. Cuthbert and St. Aratack along the Demerara River, the people are depleting whole populations of the trees (known there as "ite") by cutting them down for basket materials and other crafts. Approximately 800 trees per month are killed by a community of only 900 people. It thus seems that a nursery program and reforestation of the plant will inevitably locally ensue for this now widespread species.

In the island of Dominica, I recently introduced two plants: *Phenakospermum guyannense* (technically, perhaps "reintroduced" for it was grown in the Roseau Botanic Garden in British colonial days and has long since disappeared), and the white-flowered *Strelitzia nicolai*—the strelitzias are fruiting prolifically on the seasonally dry west coast at a hillside plot of semideciduous scrub that I acquired in 1969. The bat-pollinated *Phenakospermum* was grown from seed collected by me in 1985 at the flower-farm of Mr. Adrian Thompson near Timehri, Guyana, during a trip made under the auspices of the Smithsonian Biological Diversity of the Guianas Program; it is now planted at the Bill Harris residence near Mero, Dominica, and is still in the vegetative stage of development.

Other individual phenakospermums raised from seed collected during my 1985 trip have been variously used to establish the chromosome number of the plant (by Dr. Peter Goldblatt of the Missouri Botanical Garden, St. Louis), as material grown for experimental purposes at the National Museum of Natural History and the National Botanical Garden in Washington, D.C. by Dr. John Kress of the Smithsonian's Department of Botany, and donated to the Rainforest Habitat plantings at the National Aquarium in Baltimore, Maryland. Using such examples, more recognition and exposure of these forest plants will be sought to examine their role in the wise utilization of Guianan biological diversity, as well as in solving scientific questions that may impinge on their conservation in the future. The twenty-first century will know that numerous botanists, ethnobotanists, economic botanists, foresters, conservationists, horticulturists, agriculturists, scientists, educators, government administrators, and citizens of the present time were linked by their common concern for the well-being of the forest plants of the Guianas.

Acknowledgments

I wish to thank the late Mr. Adrian Thompson (Kitty, Georgetown, Guyana) for access to his collection of nineteenth-century Georgetown

Botanic Garden reports; Ms. Shirley Maina (Plant Conservation Unit, Department of Botany, Smithsonian Institution) for data from the National Institutes of Health workshop; Ms. Susan Richardson (Department of Botany, Smithsonian Institution) for assistance with the propagation and distribution of phenakospermum plants; and Mr. Charles Haren (PLENTY, Davis, California) for data on the condition of ita palm in parts of Guyana.

References

Anonymous. 1962. A Surinam Portfolio. *Natural History* 71(10):28–41.

Bernardi, L. 1976. J.-B. C. Fusee-Aublet, le brave botaniste de la onzieme heure. *Musees de Geneve* 169:2–10.

Bernhardt, P. 1987. Trees of Two Seasons. *Garden* 11(1):14–17.

Berry, F., and W. J. Kress. 1991. *Heliconia—An Identification Guide*. Washington and London: Smithsonian Institution Press.

Bharucha, F. R., and V. M. Meher-Homji. 1965. On the Floral Elements of the Semi-Arid Zones of India and Their Ecological Significance. *The New Phytologist* 64(2):330–42.

Brinkman, J. 1980. Surinaamse Planten in Nederland in de Zeventiende Eeuw. Unpublished Ph.D. thesis. Utrecht, The Netherlands: Instituut voor Geschiednis van de Biologie.

Brown, W. H., ed. 1920–1921. *Minor Products of Philippine Forests*. Vol. I. (1920); Vol. II, (1921); Vol. III, (1921). Manila: Bureau of Printing.

David, Y. 1966. *The Malmaison*. Paris: Le Temps.

Duval, M. 1982. *The King's Garden*. Charlottesville: University Press of Virginia.

Engle, M. M. 1988. Maria Sibylla Merian—Seventeenth Century Jungle Scientist. *South American Explorer* 17:4–11.

Glacken, C. J. 1960. Count Buffon on Cultural Changes of the Physical Environment. *Annals of the Association of American Geographers* 50(1):1–21.

de Granville, J. -J. 1988. Phytogeographical Characteristics of the Guianan Forests. *Taxon* 37(3):578–94.

Heniger, J. 1987. Surinam—Introduction of Plants from Elsewhere. Unpublished paper. Utrecht, The Netherlands: Biohistorisch Instituut, Rijksuniversiteit Utrecht.

Hills, T. L. 1961. The Interior of British Guiana and the Myth of El Dorado. *Canadian Geographer* 5(2):30–43.

Hills, T. L. 1965. *Savannas: A Review of a Major Research Problem in Tropical Geography*. Office of Naval Research. Technical Report No. 3 (unpaginated), also published as McGill University Savanna Research Project, Savanna Research Series No. 3.

Hix, J. 1974. *The Glass House*. Cambridge: MIT Press.

Honour, H. 1975. *The New Golden Land: European Images of America from the Discoveries to the Present Time.* New York: Pantheon Books (Random House).

Howard, R. A. 1953. Botanical Gardens in West Indies History. *The Garden Journal* (New York) 3:117–20.

Howard, R. A. 1954. A History of the Botanic Garden of St. Vincent, British West Indies. *Geographical Review* 44(3):381–93.

Howard, R. A. 1983. The Plates of Aublet's *Histoire des Plantes de la Guiane Françoise. Journal of the Arnold Arboretum* 64(2):255–92.

Jenman, G. S. 1888. *British Guiana. Report on the Botanic Gardens for the Year 1887.* Georgetown, Demerara.

Kingdon-Ward, F. 1956. Cosmopolitan Tropical Trees. *The Geographical Magazine* 28(10):478–88.

Koningsberger, H. 1967. *The World of Vermeer, 1622–1675.* New York: Time, Inc.

Krukoff, B. A., and H. N. Moldenke. 1938. Studies of American Menispermaceae, with Special Reference to Species Used in Preparation of Arrow-Poison. *Brittonia* 3(1):1–74.

Lacey, R. 1973. *Sir Walter Raleigh.* New York: Atheneum.

Lamb, C. 1991. Knight to Empress. *The Garden* (RHS) 116(2):71–75.

Lewis, T. A. 1989. Daniel Janzen's Dry Idea. *International Wildlife* 19(11):30–36.

Lindeman, J. C. and S. A. Mori. 1989. The Guianas, in Campbell, D. G. and H. D. Hammond, eds., *Floristic Inventory of Tropical Countries: The Status of Plant Systematics, Collections, and Vegetation, plus Recommendations for the Future.* New York: The New York Botanical Garden.

Ly-Tio-Fane, M. 1982. Contacts between Schonbrunn and the Jardin du Roi at Isle de France (Mauritius) in the Eighteenth Century. *Mitteilungen des Osterreichischen Staatsarchivs* 35:85–109.

Mauguin, G. 1933. Une imperatrice botaniste. *Organe de l'Institute Napoleon* 37:234–47.

McIntyre, A. R. 1947. *Curare: Its History, Nature, and Clinical Use.* Chicago: University of Chicago Press.

Meher-Homji, V. M. 1964. Drought: Its Ecological Definition and Phytogeographic Significance. I. Ecological definition. *Tropical Ecology* 5:17–31.

Meher-Homji, V. M. 1965. Drought: Its Ecological Definition and Phytogeographic Significance. II. Phytogeographic Significance of Drought. *Tropical Ecology* 6:19–33.

Meher-Homji, V. M. 1971. Analagous Bioclimates and Introduction of Economic Exotics. *Journal of the Bombay Natural History Society* 67(3):398–413.

Meher-Homji, V. M. 1971. On the Leaf-Fall in the Tropics. *Myforest* (Mysore, India):31–35.

Perlin, J. 1989. *A Forest Journey: The Role of Wood in the Development of Civilization.* New York and London: W. W. Norton & Company.

Plotkin, M. J., B. Boom, and M. Allison. 1991. *The Ethnobotany of Aublet's Histoire Des Plantes De La Guiane Françoise (1775).* St. Louis City: Missouri Botanical Garden.

Rodway, J. 1912. *Guiana: British, Dutch, and French.* London and Leipzig: T. Fisher Unwin.

Rowse, A. L. 1962. *Sir Walter Raleigh: His Family and Private Life.* New York: Harper & Brothers.

Schama, S. 1987. *The Embarassment of Riches: An Interpretation of Dutch Culture in the Golden Age.* New York: Knopf. (See also Book Review by Steiner, G., September 14, 1987. *The New Yorker* 63(30):130–134.)

Schnell, R. 1965. Apercu preliminaire sur la phytogeographie de la Guyane. *Adansonia* 5(3):309–55.

Schultes, R. E., and R. F. Raffauf. 1990. *The Healing Forest: Medicinal and Toxic Plants of the Northwest Amazonia.* Portland, OR: Dioscorides Press.

Stearn, W. T. 1976. The earliest European Acquaintance with Tropical Vegetation. *Gardens' Bulletin, Singapore* 29:13–18.

Stearn, W. T. 1982. Maria Sibylla Merian (1647–1717) As a Botanical Artist. *Taxon* 31(3):529–34.

Stearn, W. T. 1988. Tropical Botany and Horticulture in Europe. *Symbolae Botanicae Upsalienses* 28(3):40–47.

Stearn, W. T. 1988. Carl Linnaeus's Acquaintance with Tropical Plants. *Taxon* 37(3):776–81.

Stearn, W. T., and L. H. J. Williams. 1957. Martin's French Guiana Plants and Rudge's "Plantarum Guianae Rariorum Icones." *Bulletin du Jardin Botanique de l'Etat a Bruxelles* 27(2):243–65.

Thomas, K. B. 1963. *Curare: Its History and Usage.* Philadelphia and Montreal: J. B. Lippincott Company.

Toogood, A. 1983. Plant Trek to the Lost World. *The Garden* (RHS) 108(4):133–36.

Uichanco, L. B. 1969. Response of Some Plant and Animal Species to Physical Stress in the Tropical Environment. *Philippine Journal of Science* 98(2):155–68.

Van Lier, R. A., ed. 1972. (Stedman, J.G.) *Narrative of an Expedition Against the Revolted Negroes in Surinam.* Amherst: University of Massachusetts Press.

Vermeulen, F. E. 1985. *Preliminary Chronological Survey of Plant Collectors in French Guiana.* Utrecht, The Netherlands.

Waldman, M. 1950. *Sir Walter Raleigh.* London: Collins.

Welebit, D. 1988. The "Wondrous Transformations" of Maria Sibylla Merian. *Garden* 12(2):10–13.

Wijnands, D. O. 1983. *The Botany of the Commelins.* Rotterdam: A. A. Balkema.

Wijnands, D. O. 1987. The Hortus Medicus Amstelodamus—Its role in Shaping Taxonomy and Horticulture. *The Kew Magazine* 4(2):78–91.

Winton, J. 1975. *Sir Walter Raleigh.* New York: Coward, McCann & Geoghegan, Inc.

Zarucchi, J. L. 1984. The Treatment of Aublet's Generic Names by His Contemporaries and by Present-Day Taxonomists. *Journal of the Arnold Arboretum* 65(2):215–42.

A Selected Bibliography of Plant Conservation in the Guianas

Boxman, O., N. R. de Graaf, J. Hendrison, W. B. J. Jonkers, R. L. H. Poels, P. Schmidt, and R. Tjon Lim Sang. 1987. Forest Land Use in Suriname, in van Beusekom, C., C. van Goor, and P. Schmidt, eds., *Wise Utilization of Tropical Rain Forest Lands*. Tropenbos Scientific Series 1.

Clavel, P., J. -P. Profizi, and B. Sallee. 1978. *Comment eviter le massacre de la foret Guyanaise? Suggestions pour un develloppement de la Guyane Française fonde sur la conservation de sa foret*. Montpellier, France: Laboratoire de Botanique, Institut Botanique.

de Granville, J.-J. 1975. *Projets de Reserves Botaniques et Forestieres en Guyane*. Cayenne, French Guiana: ORSTOM.

Mittermeier, R. A., S. J. Malone, M. J. Plotkin, F. Baal, K. Mohadin, J. MacKnight, M. Werkhoven, and T. B. Werner. 1990. *Conservation Action Plan for Suriname*. Conservation International, Suriname Forest Service, World Wildlife Fund, Foundation for Nature Preservation in Suriname (STINASU), and University of Suriname.

Pearce, F. 1990. Guyana Welcomes Gene Prospectors. BBC *Wildlife* 8(8):554.

Sastre, C. 1980. Fragilite des ecosystemes guyanais: quelques examples. *Adansonia* 2, 19(4):435–49.

Sullivan, F. 1990. Proactive conservation in Guyana. WWF Reports (August/September):10–12.

Wencelius, F. 1985. The Forest Ecosystem in French Guiana: Ecological Study on Its Evolution under Man-Made Changes, in Lugo, A. E. and S. Brown, eds., *Watershed Management in the Caribbean*. Rio Piedras, Puerto Rico: Institute of Tropical Forestry.

11

Nontimber Forest Products of the Peruvian Amazon

Antonio Brack
Amazonia Associación, Peru

In the sixteenth century, Spanish conquistadors penetrated the Amazon River Basin for the first time. Inspired by the legend of El Dorado, the mythical country of sumptuous riches, they came in search of gold and cinnamon. Naturally, they found neither gold nor cinnamon, which still existed only in Asia.

Over the centuries, the legend of El Dorado changed in focus. Beginning in the nineteenth century, Peru promoted the occupation of the Amazon from the viewpoint of, first, the tremendous opportunities to obtain rubber, and second, the endeavor to develop agriculture and livestock, extract timber, and transform the jungle into "the warehouse of the country."

On the basis of these endeavors, the natives dwelling in the region were assaulted, rubber and cascarilla (or sweetwood bark) were plundered, millions of hectares of forest were cut down to make way for fields and pastures, and hundreds of millions of dollars in international loans were expended.

The results have become catastrophic. Entire valleys are devoted to coca cultivation and narcotics traffic, with all the attendant series of national and international problems. Eight and a half million hectares of forest have been cut down, 6.5 million of which have been abandoned because of the loss in soil fertility. Of the remaining 2 million some hectares have very low levels of forest, agriculture, and silviculture production in general; and of this area, 300,000 hectares are devoted to cocaine cultivation.

After almost five centuries of blunders and bankrupt development models in the Amazon, we are faced with important questions: Can we continue along the same road? Are there not other possibilities for development in the Amazon? What have been the mistakes that have led to the current situation of ruin?

These questions are quite relevant, as much for Peru as for other countries, as much for the governments as for those who sincerely strive for civilization's harmonious development in the environment of the region, as much for the international agencies that finance development projects as for those who suffer the consequences of an onerous debt that has failed to produce the fruits that were envisioned. The ecologists, the conservationists, the anthropologists, and the planners must also seek an answer to these questions. Many conservationsists are concerned because the protected areas of the region are under assault by oil exploration activities and by the migration of impoverished rural farmers fleeing poverty in the Andes who come to find land.

This chapter is an attempt to focus on certain answers on the basis of the results of the most recent research in the fields of ecology, anthropology, technology, and economics.

BIODIVERSITY HOLDS THE KEY TO EL DORADO

Biodiversity and Development

Research in recent decades has shown quite clearly that the forests and waters of the Amazon region shelter great diversity of flora and fauna species and that many of these are of utility for diverse purposes.

Both the conquistadors and the proponents of development failed to recognize that the true wealth of the area, the true El Dorado, is in the forests and the waters. For a long time, it was firmly believed that the great diversity of flora and fauna was an obstacle to the development of economically profitable activities and that it was necessary to cut down and burn the forest in order to achieve true development. On the basis of this belief and of the inappropriate technologies that were introduced, vast tracts of forest were converted into pasture, farmland, and homogenous forest.

The studies by ecologists, anthropologists, and economists in recent decades shed light on the true situation of the region by demonstrating that biodiversity is the basis for development. Ecologists have demonstrated the tremendous diversity of species within the forests and waters. Anthropologists have helped to discover that the forest systems of the

Aborigines, successfully proven over thousands of years, hold the key to the rational management of resources and to achieving harmony between humans and their environment, so that we may at the same time live in the environment and live from it in a sustainable manner.

Useful Biodiversity

Over the last three decades, knowledge has developed on the useful species of the Amazon region and the prospects for their sustainable use. Here I wish to present some data on the useful biodiversity of the Peruvian Amazon region.

Useful Species of Flora

Most of the useful flora from the Peruvian Amazon region are known to the aboriginal cultures, but have not been submitted to modern scientific experimentation. Many of them have been botanically classified, while others are known only by their denominations in native languages.

- 36 plants are oil and wax producing species, 34 of which are classified, and are known to produce edible, combustible, and industrial oils, waxes, and other similar products.
- 90 plants are known as species used for adornment or ornament, 84 of which are classified.
- 11 species are known to be used for ceramic pottery, 10 of which are classified.
- 100 species are known to be used for utensils, 77 of which are classified. These species are used to make harpoons, rafts, canoes, blow guns, darts, cabinetry, balls, large bowls and basins, arrowheads, paddles, toys, hammocks, brooms, sponges, mats, fiber, arrows, and rope.
- 4 plants are known to be species producing varnish and tar, of which three are classified.
- 35 plants are used for beverages, of which 30 are classified.
- 66 plants are known to be species used as amulets, agents of attraction, and good luck charms, 33 of which are classified.
- 17 plants are known for their properties as tanning agents.
- 526 edible plants are known, 446 of which are classified. They are used as food for their bulbs, shoots or hearts, starch, flowers, fruits, flour, leaves, fungus, latex, nut, pulp, root, seed, sepal, tubers, pods, and stalks.

• 18 plants produce saponins, of which 15 are classified. (Saponins are used as detergents or soaps.)

• 12 plants are known as species used as forage. They are used as forage, hay, and grass.

• 19 species are used for smoking and incense, 14 of which are classified.

• Several species of the Cecropia genus are used for paper.

• 401 plants are known to produce wood, of which 318 are classified.

• 22 species are known that are particularly appreciated for charcoal, firewood, and light providers.

• 8 species are known as caustics.

• 21 species are known to produce rubber, copal, chewing gum, gutta-percha, resin, latex, and other latex products.

• 3 species produce gum.

• 8 species are appreciated for hedge and shade.

• 4 species are used to produce fabrics.

• 110 species are known to be used as coloring agents and dyes, of which 65 are classified.

• 22 species are used as condiments.

• 41 species are appreciated for construction in support beams and roofing material.

• 334 species are used as toxics, of which 308 are classified. They are used for curare, anti-tick and chigger agents, herbicides, ichthyo-toxins, insecticide, agents to prevent parasites, poisons, toxins, and other related materials.

• Over 2,000 plants in the Peruvian Amazon region have been recorded by different authors as used for curative purposes; many of them have yet to be classified.

Domestic Species and Plant Varieties

Over 100 domestic species of native flora are known in the Peruvian Amazon region. Some of them have dozens of varieties. These constitute an important germ plasm source that has yet to be studied much in the Peruvian Amazon.

Useful Species of Fauna

The fauna of the Peruvian Amazon region are widely used to obtain foods (meat), furs, hides, adornments, and so on. Over 100 species of mammals, birds, reptiles amphibians, fish, mollusks, insects, and other groups are used for these purposes.

BIODIVERSITY AND ECONOMICS

The most recent research is furnishing very interesting data on the total economic value of the Amazon ecosystems. The value of direct use continues to increase, not only the use of forest timber, but also the useful biodiversity, that is, flora and fauna species. The direct use of these different nontimber products is a profitable activity and one of great importance in the Peruvian Amazon region, mainly in the supply of food products at the local and regional levels.

Biodiversity and Nutrition

Hydrobiological resources, especially fish, constitute a very important source of protein in the region. Fish consumption increases annually at 30,000 metric tons, and in the city of Iquitos alone, some 13,000 metric tons of fish are consumed each year. This is the most important source of protein in the region, surpassing all other sources combined.

Hunting of terrestrial fauna (known as "mountain meat") is the second most important source of protein in the region, reaching some 8,000 metric tons of consumption a year.

Total production of beef, fowl, pork, barnyard-raised meat, milk, and eggs grew to 17,574 metric tons just in the Department of Loreto in the Amazon region. These data, when compared to fish and terrestrial meat production, are quite revealing if one considers that livestock development has cost Peru some $500 million (U.S. dollars) over the last 50 years, and that millions of hectares of forests have been cleared for this purpose.

Harvest of Forest Products

The harvest of forest products in a comprehensive fashion is an important economic activity in the Peruvian Amazon region and more profitable than many forest production systems based on the slash-and-burn method.

In Tamshiyacu, near Iquitos, some 2,000 people work in the collection of products and rotational crop farming to supply themselves and the market at Iquitos. Seven of their main productive activities, in descending order of importance are fruit farming (63 percent), intensive agriculture (21 percent), fauna products (9 percent), charcoal (3 percent), fibers and handicrafts (2 percent), wild fruits (1 percent), and medicinal plants (0.5 percent). The system is sustainable due to a long rotation period of up to 30 years between farming and forest.

In Mishana, on the Nanay River, near Iquitos, a group of 50 river families make use of a variety of forest products that they market in Iquitos. One hectare of forest contains 275 species of trees and 842 individual tree species. Of this total number, 72 species and 350 individuals are sources of marketable products: 11 species produce edible fruits, 60 species produce commercial lumber, and 1 species, the shiringa, produces rubber. The river dwellers of Mishana, in addition to collecting forest products, also fish, hunt, and rotate crops that they farm on small plots of land. The value they receive for total production per hectare is about $1,700 for product collection, a figure that is quite respectable in comparison to those obtained by livestock and agriculture.

A preliminary economic analysis indicates that livestock production on 200 hectares of central Peruvian jungle incurs losses. Total investment for a ten-year period in forest clearing, livestock inputs, and pasture maintenance runs $49,000. The value of losses in natural forest productivity (timber, foodstocks, and rubber) reaches $3,694,000, whereas total production value of livestock and fish farming is $142,000 over the same period. Thus, over the ten-year period, annual production per hectare incurs a net loss of $1,800.

An analysis of the consumption patterns in important cities in the Peruvian Amazon region clearly establishes a high nutritional dependency on the harvest of resources from aquatic and forest ecosystems. For example, in the year 1988, the city of Iquitos consumed no more than 500 metric tons of beef, but 13,000 metric tons of fish and a large amount of native fruits harvested from nearby forests.

SOME OPTIONS FOR THE FUTURE

There are some very important points to consider in the future development of the Amazon. These points should also be of serious concern to international organzations, both those that depend on government assistance and those that do not.

1. Studies on Useful Biodiversity and Its Economic Importance at the Local, Regional, and Global Level

It is imperative that more information be obtained on useful biodiversity, production in ecosystems, and the economically profitable management of forest and hydrobiological resources that utilize low-impact and alternative technologies.

In order to persuade politicians and planners, there is a need for objective and convincing data that compare the systems based on forest clearing with those based on the management of natural systems. This data should focus on several factors. For example, national data bases that collect existing information and generate new data on useful species should be considered. Clearinghouses of information and documentation from research and investigation should be established that would make information accessible to organizations and interested native citizens. At the present time, information must be repatriated; in the instance of biodiversity in Peru, about 90 percent of the information can only be found abroad. Research on useful plants and animals, from taxonomies and characteristics to marketing opportunities, must be encouraged. Economic studies of traditional systems of regional resource management that admit comparison with implanted systems and emphasize feasibility must be conducted. These economic studies should focus on the aggregate economic value of Amazon ecosystems, concentrating on their direct and indirect value, including opportunity costs, for the value of inevitable use.

2. Land Management

It is of fundamental importance to undertake land management in the Amazon in order to determine area mangement options. Generally, construction of roads and the establishment of settlements have occurred without allocating space on the basis of land–use capacity, spatial occupation (aboriginal communities), or the location of biodiversity.

The kind of land management that must be considered would mark off areas of different management alternatives into protective zones of various classes (parks, preserves, and so on), areas of comprehensive and permanent forest management (wood as well as other products), areas of hydrobiological resource management, which could be developed in combination with the aforementioned areas, already colonized areas, and lands of traditional resource use of aboriginal groups.

The establishment of areas for the traditional use of natural resources by aboriginal groups in the form of community reserves, extractive reserves, and so on is very important in order to conserve the culture, technologies, and useful biodiversity in situ. The conservation of the useful biodiversity of the Amazon can be fully accomplished only in concert with the conservation of aboriginal cultures.

3. Improving Production in Previously Colonized Areas

Undertaking activities to improve the efficiency of production systems in previously colonized areas is another priority objective. There are millions of deforested areas that are either abandoned or that yield low levels of production. It is feasible to improve and manage these areas by using systems to increase production and keep pressure from unspoiled areas.

Priorities in this regard should focus on management of secondary growth forests, where there is great productive potential for timber, fauna, apiculture, and other areas, and on implementation of technologies that guarantee sustainable production using, agroforest systems, heterogenous and permanent native crops, and flora collection. Priorities should also focus on multidisciplinary work. It is important both to work with greater reliance on multidisciplinary teams (ecologists, economists, anthropologists, and technologists) as well as to focus the research itself on a more comprehensive and multidisciplinary perspective. There are quite a few people who believe in the great biodiversity of the Amazon region, but few have understood the tremendous economic utility of the array of species for the future of the region.

References

Bodmer, R. et al. 1988. Ungulate Managment and Conservation in the Peruvian Amazon. *Biol. Conserv.* 45:303–10.

Brack, A. J. Ocaña et al. 1990. *Desarrollo Sostenido de la Selva.* INADE-APODESA, Lima.

Denevan M., and C. Padoch. 1990. *Agroforesteria tradiconal en la Amazonia peruana.* CIPA, Documento 11, Lima.

Hiraoka, M. 1985. Floodplain Farming in the Peruvian Amazon. *Geog. Rev.* (Ser. B) 58(1):1–23.

Hiraoka, M. 1986. Zonation of Mestizo Riverine Farming Systems in Northeast Peru. *Nat. Geogr. Research.* 2(3):354–71.

Neyra, R. et al. 1991. *Informe sobre Desarollo y Medio Ambiente en el Sector Pesquero.* Ministerio de Pesquería, CNUMAD 1992.

NRC, 1982. *Ecological Aspects of Development in the Humid Tropics.* Washington, D.C.: National Academy Press.

Oldfield, M., and J. Alcorn. 1987. Conservation of Traditional Agrecosystems. *BioScience* 37(3):199–208.

Padoch, C. et al. 1985. Amazonian Agroforestry: A Market-Oriented System in Peru. *Agroforestry Systems* 3:47–58.

Pearce, D. 1991. Deforesting the Amazon: Toward an Economic Solution. *Ecodecision* 1:40–49.

Peters, C., A. Gentry, and R. Mendelsohn. 1989. Valuation of an Amazonian Rainforest. *Nature* 339:655–56.

Rutter, A. 1990. *Catálogo de las Plantas Utiles de la Amazonia Peruana.* Min. De Educ. Instituto Linguistico de Verano. Lima.

Soukup, J. 1988. *Vocabulario de los nombres vulgares de la flora peruana y Catálogo de Géneros*. Salesiana. Lima.

Vásquez, R., and A. Gentry. 1989. Use and Misuse of Forest-Harvested Fruits in the Iquitos Area. *Conserv. Biol.* 3(4):1–12.

Webb, R., and G. Fernández Baca. 1990. *Perú en Números*. Almanaque Estadístico. Cuanto S.A., Lima.

12

Products from the Tropical Rain Forests of Mexico: An Ethnoecological Approach

VÍCTOR M. TOLEDO, ANA I. BATIS, ROSALBA BECERRA,
ESTEBAN MARTÍNEZ, AND CLARA H. RAMOS
Centro de Ecología, Universidad Nacional Autónoma de México

After several centuries, Western civilization continues to utilize tropical rain forests in a highly destructive manner, as can be measured by the annual rates of deforestation, which are estimated to have reached 7.15 million hectares from 1981 to 1985 (Lanly, 1984). Furthermore, this rate appears to be on the rise (World Resources Institute, 1990). Such deforestation results from utilization schemes that either clear forests to introduce crops, timber plantations, and cattle pasture, or makes inefficient use of standing forest products, exploiting only a few species of trees as timber.

In response to this situation, a growing number of researchers (anthropologists, biologists, geographers, and human ecologists) have been carrying out studies on the indigenous methods of using and managing tropical moist forests. These studies are revealing not only the large number of species of plants, fungi, and animals used by the indigenous cultures of tropical moist areas, but also the ecological feasibility of their management strategies (Posey et al., 1983; Clay, 1988; Gómez-Pompa et al., 1991). This *ethnoecological approach* is perhaps the most appropriate means for attaining sustainable use of tropical ecosystems. However, very few studies have been done to demonstrate the actual economic feasibility of indigenous use and management of tropical natural resources.

This chapter presents data on the total number of useful plants and plant products found in the tropical moist forests (primary and secondary) of Mexico based on the knowledge of several indigenous groups, and discusses the economic importance of the forests, as well as their significance for conservation.

THE ETHNOECOLOGICAL ALTERNATIVE: PREVIOUS STUDIES

The whole spectrum of useful (plant and animal) species of a micro-region with natural tropical forest has only been calculated twice. The first area was the region of Uxpanapa (Veracruz) in Mexico, where an area of approximately 1,200 hectares of tropical rain forest included over 1,000 species of plants and vertebrate animals. Based on the ethnobotanical and ethnozoological information from 13 studies, and on the knowledge of natives, the traditional uses of each species existing in the Uxpanapa region was inventoried. As a result, over 700 useful products were obtained (Toledo et al., 1978; Toledo, 1991b). In Africa, Richards (1991) developed a similar (although more ethnographic) research project in the Gola Forest of Sierra Leone, where 90 households of three indigenous communities obtained 1,355 items from primary and secondary forests, rivers, streams, farms, and plantations. On the other hand, some studies have provided data on the products obtained from tropical forests at the site level (one hectare or less) (Prance et al., 1987). In these cases, the percentage of useful tree species per hectare ranged from 78.7 percent and 76.8 percent among the Chacobo of Bolivia and Ka'apor of Brazil, to 61.3 percent and 48.6 percent among the Tembe of Brazil and Panare of Venezuela (Prance et al., 1987). Finally, the market value of the biological resources obtainable from a given area in a natural tropical forest has also been calculated by some authors. For example, Peters et al. (1987) estimated the net present value of the tree products in a one-hectare plot of the Mishana forest in the Peruvian Amazonia, and Alcorn (1989) made what is perhaps the first economic evaluation of the entire productive system of an indigenous group while working with the Huastec Indians of Mexico.

THE TROPICAL FOREST ZONES OF MEXICO

In Mexico the tropical moist zone, characterized by an original covering of medium and high forests and/or savannahs, extends over nine southern and southeastern states. Estimated by Rzedowski (1978) to cover approximately 11 percent of Mexico, this zone is comprised of 20.15

million hectares distributed in 251 municipalities (Toledo and Ordóñez, 1991). These regions have been severely degraded by farming that has produced a deforested area equivalent to 40 percent of the total by the end of the 1970s and early 1980s; the annual deforestation rate at this time is 237,000 hectares (Massera et al., 1991). As a result, there are only six regions left with primary forest: Los Tuxtlas, Veracruz (40,000 hectares), Uxpanapa-Chimalapas in Veracruz and Oaxaca (nearly 300,000 hectares), the Lacandon Forest of Chiapas (330,000 hectares), Calakmul in Campeche, and the Sian Ka'an Reserve in Quintana Roo (145,000 hectares). From the point of view of biodiversity, the tropical rain forests of Mexico are, like their counterparts in the rest of the world, the most species-diverse type of vegetation in the country. The Lacandon Forest in Chiapas, for example, contains around 4,000 species of vascular plants (E. Martínez, unpublished), 130 species of mammals, over 300 species of birds, and nearly 100 species of reptiles and amphibians (Vega-Rivera, 1990).

INDIGENOUS PEOPLES OF THE TROPICAL FORESTS OF MEXICO

Eighteen indigenous groups live in the tropical moist zones of Mexico. With a population of over 900,000 (Argueta, 1990), the indigenous cultures living in these areas can be divided into three well-defined groups: 1) those groups whose main habitat is in this zone (Chinantecs, Choles, Chontales, Huastecs, Lacandons, and Totonacs), 2) those groups for whom the tropical moist habitat is marginal, as the bulk of their population lives in other ecological zones (for example Mixtecs, Otomis, Tepehuas, and Tzeltals), and 3) those groups such as the Nahuas and Mayas that have a significant proportion of their population in this and other zones by virtue of their large numbers. These groups generally live in rural communities working in slash-and-burn agriculture, extraction of forest products, hunting and fishing, and small-scale cattle raising. Their production systems reflect a multiple use agroforestry strategy, as in the rest of Latin America. There are some ethnic groups that have been extensively studied from the ethnobotanical standpoint, as is the case of the Huastecs (Alcorn, 1984), the Mayas (Barrera et al., 1976), and the Totonacs (Medellín, 1988). Reports with considerable information have been presented on other groups, such as the Lacandon Maya (Nigh and Nations, 1980) and the Chujes (Breedlove and Hopkins, 1970). For 7 of the groups no ethnobotanical information is available. Even though only 10 of the 18 ethnic groups have been studied from the standpoint of their botanical knowledge, the most important groups (gauged by population and/or being the exclusive and long-standing dwellers of their respective habitat) have been covered by one or more ethnobotanical studies, for example, the Totonacs and Lacandon Maya.

THE CONCEPTUAL FRAMEWORK: THE INDIGENOUS MODEL IN THE TROPICAL MOIST ZONES

The data presented in this article are from the research done in a more extensive project aimed at evaluating the indigenous model in the moist tropics from the ecological and economic standpoint. Until very recently, studies geared to determining the indigenous groups' knowledge and ways of using nature were guided either by a search for new resources for Western civilization, or by simple intellectual or academic curiosity. This is well illustrated by the case of ethnobotany, which in its most orthodox version was either economic botany (generally practiced by botanists) or ethnoscience (practiced by ethnographers and linguists).

In the past, ethnobotany operated as an instrument for the merely academic understanding of the role of plants in the material and non-material cultures of indigenous societies. Today, that situation has changed. The worsening of the ecological crisis on a world scale, the loss of biotic and genetic resources, deforestation in the tropics, the great advances in the indigenous struggles, and the spread of the environmental movement and the movement for democracy have shaken the consciousness of a great many researchers.[1] As a result, the old colonialist and neocolonialist approaches to research are no longer feasible. The approach to the indigenous cultures should be centered on a complete reassessment of their societal models and of their articulation with nature. This new ethnoecological research is based not so much on the search for techniques, raw materials, or forms of knowledge, but on the new paradigm that recognizes in the indigenous cultures the starting point for designing ecologically sound ways of using natural resources. It is in this context that the search for and evaluation of indigenous models of natural resource management zones of the world has become the key objective of the ethnoecological approach (Toledo, 1991a). From the abundant literature on the ecological strategies of indigenous peoples in tropical rain forest regions in Latin America, it is possible to postulate the existence of an indigenous model through which the traditional cultures have made ecologically sound use of the tropical ecosystems. This model operates like a system, a totality made up of several segments, on the basis of which the indigenous cultures carry out their productive processes. This system is but the particular expression of the native strategy for appropriating nature (Toledo, 1990). The indigenous model of the tropical moist zones of Latin America is made up of eight main segments: 1) the primary forests, 2) the planting areas, 3) the livestock areas, 4) the areas with forestry plantations, 5) the bodies of water, 6) the homegardens, 7) the secondary forests, and 8) the managed forests.

METHODS

To assist in the economic assessment of the indigenous model for the use of tropical natural resources, an information system has been used that makes it possible to take stock of the number of useful plant species and their products that have been recognized through indigenous knowledge. This information system is made up of two data banks: one is based on flora, and the other is ethnobotanical. The first covers the plant species registered by the main existing inventories of flora, as well as those derived from special collections done by the botanists working with the project. This data bank is capable of managing lists of plant species on four different scales: 1) *sites,* which include samples of 1 hectare or less, 2) *localities,* with samples of 1 hectare to 10 hectares, 3) *regions,* with inventories of areas of approximately 1,000 hectares, and 4) *national,* in other words, all species registered for Mexico. The information from this flora data bank is from two main sources: the botanical studies done by several authors in six strategic regions (Los Tuxtlas, Veracruz; Tuxtepec, Oaxaca; Uxpanapa-Chimalapas in Veracruz and Oaxaca; Sian Ka'an, Quintana Roo; and Lacandon and El Soconusco in Chiapas), and the ecological inventories taken at specific sites. Thus, the data bank is expected to include 3,000 to 4,000 vascular plant species of the total 5,000 species estimated by Rzedowski (1991) to exist in all of the flora of the tropical moist areas of Mexico.

The second data bank includes all species of the tropical forests of any utility based on the knowledge of several indigenous groups of Mexico. This data bank is being built with information from existing ethnobotanical studies (approximately 30 studies), which to date cover 11 of the 18 ethnic groups living in this ecological zone. Each species input into the data bank is accompanied by a file containing information on the scientific and common name, botanical family, location, type of use, part used, life form, habitat (primary and/or secondary forest), and the indigenous group from which the data were obtained.

From the integration of these two data banks into the ethnobotanical data bank, it is possible to determine the number of plant products that exist in the tropical moist forests of Mexico according to the knowledge of the indigenous groups. The register of the number of products can be done from a scale that includes 1-hectare sites located in different geographic locations of the country, to whole regions, and finally for all of the tropical moist flora of Mexico. This will make possible a complete analysis of the useful flora at different levels and in different geographic areas.

RESULTS

The data presented were obtained in 1991 from the data banks described in the previous section. A total of 1,380 plant species from the tropical moist zones of Mexico were identified as having one or more uses, in accordance with the knowledge recorded by the ethnobotanical study done with eight indigenous groups. Of this total, 465 are species found exclusively in primary forests, 712 are species found in secondary forests (153 of the two preceding figures are species found in both), and 356 are species without ecological specification of habitat. Thus 1,024 useful species can be found in primary and/or secondary forests. The useful species included in the data bank belong to 143 families, of which the most abundant are *Leguminosae, Compositae, Solanaceae, Euphorbiaceae,* and *Rubiaceae.* By life form, the most common species were trees and herbs, followed by shrubs; together these forms account for three-fourths of the species registered. Of all the species listed in the data bank, only 1,024 are known with certainty to live in primary or secondary forests. A total of 1,923 products of these species were registered: 969 in primary forests and 1,341 in secondary forests (with some overlap of products from species that are found in both habitats). These products were ordered based on the type of use and the part used. Finally, the specific uses of the two most common categories of use (medicinal and food) were broken down to illustrate the full variety of possible uses.

DISCUSSION

Seven main conclusions can be drawn from the above analysis:

1. There is a great variety of "ethno-products" available from the Mexican tropical forests: 1,923 products from 1,024 useful species. This figure indicates a ratio of 1.9 products per useful species, or almost two products per plant species recognized as useful by the indigenous peoples. However, since our results are only partial and preliminary, we can predict higher figures; perhaps approximately 1,500 useful species and nearly 3,000 products should be expected as the final figures for the

Mexican rain forests. Based on this assumption, it can be predicted that new contributions of a similar type in larger tropical countries (for example, Indonesia or Brazil) will yield more impressive results.

2. The tropical forests are important as a primary resource for three types of products: medicines (642 products, or 33 percent of the total), foods (318, or 16.5 percent), and woods and wood products (lumber, fibers, and others) (286, or 15 percent). The remaining 36 percent of products was distributed among domestic items, fuel woods, drugs, ornamentals, forages, and other goods such as resins, dyes, gums, tannins, flavorings, sweeteners, and work tools.

3. There are many more nonwood than wood products (83 percent versus 17 percent). This finding is counter to the idea that tropical forests are predominantly timber sources, and confirms the fact, recently stressed by Gentry and Dodson (1987), that tropical forests contain fewer trees than nontree species.

4. Secondary forests are extremely important as potential reservoirs of products, in this analysis yielding a higher number of plant products than the primary forests (1,341 as compared to 969).

5. The importance of primary and secondary forests is different: while primary forests provide mainly wood materials and foods, secondary forests, which also produce foods, are basically an extensive storehouse of substances such as medicines, drugs, stimulants, tannins, waxes, and dyes.

6. Products whose extraction are nondestructive (those whose source is in the form of leaves, fruits, bark, seeds, flowers, or exudates) are more abundant than products whose extraction involve a destructive action (wood, trunks, roots, tubers, or the entire plant), suggesting that sustainable use of tropical rain forests is possible.

7. While most of the food products come from edible fruits and seeds, proving that humans take advantage of tropical plant adaptations to dispersal by vertebrate animals (Oldfield, 1981), leaves (and entire plants of secondary forests) provide mainly medicines and other substances, while trunks, of course, provide wood.

The preliminary data presented in this chapter, which is the first quantitative ethnobotanical survey on a national scale, definitely confirm the assertion that tropical rain forests not only have a very high degree of biological diversity, but they also contain an exceptionally large number of useful species. At the same time, the methods used in the study to determine this economic potential—thus the knowledge accumulated by traditional peoples—confirm the essential role to be played by tropical indigenous cultures in the sustainable use of these ecosystems. Although the (ethno) products presented here are subsistence goods for domestic or microregional consumption, they represent an important source of marketable products in an ecologically oriented society. The results are certainly impressive, and should be useful in defending both tropical forests and tropical peoples, and in promoting a new and much needed green economy. This is our hope and our desire.

Acknowledgments

The authors are most grateful to the following colleagues for providing unpublished information: R. Nigh, M. A. Martínez-Alfaro, P. Richards, O. Massera, M. de Jesús Ordóñez. We received valuable technical assistance from Carmen Solís. This work was done thanks to support from the MacArthur Foundation through the Centro de Ecología of the National Autonomous University of Mexico (UNAM).

Notes

1. The ethical and social aspects of ethnobiological and ethnobotanical research have been discussed in several recent meetings such as that organized by the University of Wisconsin (International Conflict and Cooperation: Environmental Security in the 1990s, March 9–10, 1990), the Latin American Studies Association (XVI International Congress; Washington, D.C., April 4–6, 1991), and Conservation International (see chapters by L. G. Joly, K. Moran, and D. Posey). In addition, Cultural Survival has implemented a project for the marketing of forest products from the Amazon region based on the principles of ecological economics (see chapter by J. Clay).

References

Alcorn, J. B. 1984. *Huastec Mayan Ethnobotany*. Austin: University of Texas Press.

Alcorn, J. B. 1989. An Economic Analysis of Huastec Mayan Forest Management. In Browder, J. E. (ed.), *Fragile Lands of Latin America*. Boulder, Colorado: Westview Press.

Anderson, P. et al. 1984. *Estudio sobre las plantas medicinales de San Pablito, Pahuatlán, Puebla*. Informe UAM-IZTAPALAPA.

Argueta, A. et al. 1978. *Etnobotánica y uso diferenciado en una región cálido-húmeda del norte de Puebla*. Memorias del Simposio de Etnobotánica, Deas-INAH.

Argueta, A. 1990. Relaciones indigenas, proteccion ambiental y desarrollo. *El Gallo Ilustrado (El Dia)* 1462:14–16.

Argueta, A. et al. (in press). *Etnobotánica comparada: náhuas y totonacos del norte de Puebla*.

Barrera, A. M., A. V. Barrera, and R. M. Lopez Franco. 1976. Nomenclatura etnobotánica maya. Una interpretación taxonómica. INAH.

Basurto, Peña F. 1982. *Huertos familiares en dos comunidades nahuas de la Sierra Norte de Puebla: Yancuitlapan y Cuauhtlapanaloyan.* Tesis Fac. de Ciencias. UNAM.

Breedlove, D. E., and N. A. Hopkins. 1970. A Study of Chuj (Mayan) Plant Names, with Notes on Uses. *Wasman J. Biol.* 28 (part 1:275–279); 29 (part 2:107–28); 29 (part 3:189–205).

Brockway, L. H. 1979. *Science and Colonial Expansion: The Role of the Royal Botanic Gardens.* New York: Academic Press.

Browder, J. E. (ed.). 1989. *Fragile Lands in Latin America.* Boulder, Colorado: Westview Press.

Caballero. 1984. *Plantas comestibles utilizadas en la Sierra Norte de Puebla por totonacos y nahuas.* Tesis. ENEP-IZTACALA. UNAM.

Calatyud, G. A. 1990. *Estudio etnobotánico de una comunidad nahuas de la Sierra de Santa Marta.* Tesis. Profesional de Biologia. Univ. Veracruzana.

Castro Ramirez, A. E. 1988. *Estudio comparativo del conocimiento sobre plantas medicinales utilizadas por 2 grupos étnicos del municipio de Pahuatlán, Puebla.* ENEP-IZTACALA.

Clay, J. W. 1988. *Indigenous Peoples and Tropical Forests.* Boston, MA: Cultural Survival, Inc.

Espadas Resendiz M. et al. 1982. *Las plantas medicinales de una comunidad totonaca de la Sierra, Tuzamapan de Galeana, Puebla.* ENEP-IZTACALA.

Gentry, A. W., and C. Dodson. 1987. Diversity and Biogeography of Neotropical Vascular Epiphytes. *Ann. Missouri Bot. Gard.* 74:235–243.

Gómez-Pompa, A., T. C. Whitmore, and M. Hadley (eds). 1991. *Rain Forest Regeneration and Management.* MAB Series, Volume 6. Washington, D.C.: The Parthenon Pub. Group.

Gonzalez, R. C. 1989. *Estudio etnobotánico de plantas comestibles de 4 ejidos de la Sierra de Santa Marta, Veracruz.* Tesis profesional.

Hernandez, Lopez, J. A. 1988. *Estudio sobre herbolaria y medicina tradicional en el municipio de Misantla, Veracruz.* Tesis. Facultad de Ciencias UNAM.

Kelly, I., and A. Palerm. 1952. *The Tajín Totonac.* Part I. Institute of Social Anthropology. Publication No. 13.

Lanly, J. P. 1982. *Los Recursos Forestales Tropicales.* Estudios FAO-Montes 30, Rome.

Lipp, F. J. 1971. Ethnobotany of the Chinantec Indians, Oaxaca, Mexico. *Economic Botany* 25:234–44.

Martinez Alfaro, M. A. 1983. Etnomicología y exploraciones micológicas en la Sierra Norte de Puebla. *Boletín de la Sociedad Mexicana de Micología* 8:51–63.

Martinez Alfaro, M. A. 1984. Medicinal Plants Used in a Totonac Comunity of the S. N. de Puebla. *Journal of Ethnopharmacology* 11:(2):203–21.

Massera, O., M. J. Ordonez, and R. Dirzo. 1991. Carbon Emissions from Deforestation in Mexico: Current Situation and Long Term Scenarios. Report to the International Energy Studies Group, Lawrence Berkeley Laboratory.

Mata Pinzon, M. A. de La Soledad. 1988. *Estudio etnobotánico de las plantas medicinales entre los zoque-popoluca de Piedra Labrada, Veracruz.* Medio Ambiente y Comunidades Indígenas del Sureste: prácticas tradicionales de producción, rituales y manejo de recursos.

Medellín, S. 1988. *Arboricultura y silvicultura tradicional en una comunidad totonaca de la costa.* Tesis de Meastria. Ecologia y Recursos Bioticos. INIREB. Xalapa, Veracruz.

Morales Garcia, G. et al. 1987. *Contribución al estudio de la flora medicinal y medicina tradicional del municipio de Coxquihui.* Ver. Tesis. Facultad de Ciencias. UNAM.

Nations, J. D., and R. B. Nigh. 1980. The Evolutionary Potential of Lacandon Maya Sustained-Yield Tropical Forest Agriculture. *Journal Anthropological Resources* 36:1–30.

Nigh, R. B., and J. D. Nations. 1983. La Agrisilvicultura Tropical de los Lacandones de Chiapas. *Civilización* (1):341–71.

Oldfield, M. L. 1981. Tropical Deforestation and Genetic Resources Conservation. *Studies in Third World Societies* 14:277–345.

Peters, C., A. W. Gentry, and R. Mendelsohn. 1987. Valuation of a Tropical Forest in Peruvian Amazonia. *Nature* 339:655–56.

Posey, D. A. et al. 1984. Ethnoecology as Applied Anthropology in Amazonian Development. *Human Organization* 43:95–99.

Prance, G. T., W. Balee, B. M. Boom, and R. L. Carniero. 1987. Quantitative Ethnobotany and the Case for Conservation in Amazonia. *Conservation Biology* 1:296–310.

Richards, P. 1991. Saving the Rain Forest: Contested Futures in Conservation. In S. Wallman (ed.), *Anthropology and the Future.* RKP Press (in press).

Rzedowski, J. 1978. *Vegetacion de Mexico.* Limusa-Wiley.

Rzedowski, J. 1991. Diversity and Origins of the Phanerogamic Flora of Mexico. In T.P. Ramamoorthy et al. (eds.) *Biological Diversity of Mexico: Origins and Distributions.* Oxford: Oxford Univ. Press (in press).

Sanabria, O. L. 1986. *Etnoflora Yucatanense. El uso y manejo forestal en la comunidad de xul, en el sur de Yucatán.* INIREB.

Santos, R. M. A. 1988. *Etnobotánica. Plantas medicinales de los zoque-popolucas de 4 ejidos de Soteapan, Veracruz.* Tesis Profesional.

Toledo, V. M. 1990. The Ecological Rationality of Peasant Production. In M. Altieri and S. Hecht (eds.) *Agro-Ecology and Small Farm Development.* CRC Press.

Toledo, V. M. et al. 1978. El uso multiple de la selva basado en el conocimiento tradicional. *Biotica* 3:85–101.

Toledo, V. M. 1991a. *El Juego de la Supervivencia: un manual para la investigacion etnoecologica en Latinoamerica.* CLADES, Santiago, Chile.

Toledo, V. M. 1991b. Bioeconomic Costs of Transforming Tropical Forests in Pastures in Latin America. In T. Downing, S. Hecht, and H. Pearson (eds.) *Development or Destruction: The Conversion of Tropical Forest to Pastures in Latin America.* Boulder, Colorado: Westview Press (in press).

Toledo, V. M., and M. J. Ordonez. 1991. Biodiversity Scenarios of Mexico:

An Analysis of Terrestrial Habitats. In T. P. Rammamoorthy et al. (eds.) *Biological Diversity of Mexico: Origins and Distributions.* Oxford: Oxford Univ. Press (in press).

Vega-Rivera, J. H. 1990. Situacion actual del conocimiento faunistico de la Reserva Montes Azules. In J.L. Camarillo and F. Rivera (eds.) *Areas Naturales Protegidas en Mexico y Especies en Extincion.* ENEP-Iztacala, UNAM.

Villers Ruiz, M. A. et al. 1978. *Uso de maderas y otros materiales vegetales en la construcción de la habitación rural en la zona de Cobá, Quintana Roo.* Tesis. Fac. de Ciencias. UNAM.

World Resources Institute. 1990. *World Resources 1990–91.* Oxford: Oxford Univ. Press.

13

Nontimber Forest Products in the Petén, Guatemala: Why Extractive Reserves Are Critical for Both Conservation and Development

---

CONRAD REINING AND ROBERT HEINZMAN
Conservation International

Continued population growth and economic expansion in Central America, attended by poverty and social inequity, have deforested two-thirds of the region (Nations and Komer, 1984). In the department of the Petén, Guatemala, agricultural frontier expansion and logging have resulted in the conversion of between 200 and 400 square kilometers per year for the last 10 to 15 years (CATIE, 1990a, 1990b). Population has risen from 64,500 in 1973 to a present level of 250,000–300,000 (CONAP, 1989; Schwartz, 1990). Under the current pattern of extensive agriculture and cattle raising, widespread forest conversion will continue, threatening the region's social stability and ecological and economic productivity.

In response to this threat, the Maya Biosphere Reserve (MBR), encompassing nearly 1.5 million hectares, was established in the northern Petén in 1990. The Consejo Nacional de Areas Protegidas (CONAP) was established to administer this and other reserves in the country. Although the legal framework is in place for the management of the Reserve, there are many challenges facing the institutions and communities responsible for the MBR and its natural and cultural resources.

The MBR includes zones for both the strict preservation of forests, wetlands, and other natural ecosystems, as well as the sustained management of extensive forested areas (6,000 to 8,000 square kilometers). The Reserve, which occupies the northern 40 percent of the Petén, is located within a larger bioregion that includes southwestern Campeche and southeastern Quintana Roo in Mexico, as well as northwestern Belize.

The MBR is the center of an extractive industry based on nontimber forest products (NTFPs). There are three products collected from the Reserve that are widely traded internationally, namely *chicle* (Palmae), the latex of *Manilkara zapota* (Sapotaceae), *xate,* the leaves of two species of *Chamaedorea* palm, and *pimienta gorda* or allspice, the dried berry of *Pimienta dioica* (Myrtaceae). These species are common forest plants and are widely collected full- or part-time by thousands of people, many of whom depend on these resources for a substantial part of their cash income. Furthermore, the extraction of these products has been practiced for at least 30 years in the case of *xate* and *pimienta gorda,* and more than 90 years in the case of *chicle.* Therefore, these resources are exploited by a well-established forest culture, with considerable knowledge and experience (Schwartz, 1990).

Xate (Chamaedorea spp.) palms are harvested wild in the tropical forest of the Guatemalan Petén. (See Figure 13-1 for the form and relative size of the two most commonly harvested species.) The harvesters clip one or two leaves off each palm as they walk through the forest, but leave the palm plant alive so that they can return to the same plants three or four months later and clip off more leaves. Heinzman et al. (1990) and Reining et al. (1991) indicate that this cycle can be repeated indefinitely if sound harvesting practices are followed. The harvested palm leaves are exported to the United States and Europe, where they serve as the green backdrop for cut flower arrangements.

Chicle is one of the natural latexes used in the manufacture of chewing gum bases. It is harvested from *Manilkara zapota,* by *chicleros* who scale the trees, hacking zig-zagging channels in the bark. The raw latex is coagulated by the chicleros and formed into 10-kilogram *marquetas* stamped with the chicleros' initials (Figure 13-2). While chicle has been extracted throughout Central America, the largest concentration of high grade chicle is found in the Maya Bioregion. Though relatively expensive at $1.70 per pound (1990–91), chicle is a high quality latex. However, with the advent of cheap synthetic gums, rising labor costs, and cheaper natural substitutes, demand for the latex of this common forest tree declined, leaving the industry nearly dormant in the early 1980s. Nonetheless, with the recent interest in natural products, especially those produced sustainably from tropical forests, the market for chicle has rebounded considerably. Virtually all buyers at this point are Japanese.

Allspice is the berry of another common forest tree, *Pimienta dioica.*

FIGURE 13-1 The two most commonly harvested species of *xate* palm, *Chamaedorea elegans* and *C. oblongata*. *C. elegans,* the smaller species with finer leaves, is commonly called *xate hembra*. The larger species, *C. oblongata,* is called *jade* or *xate macho*.

Whole families are involved in climbing the trees, cutting and harvesting the fruit-bearing branches, and drying the green berries. Besides being used as a condiment, allspice is also used to preserve fish in European countries and as a flavoring and curing agent in processed meats and bakery products. While plantations in Jamaica supply most of the world's allspice, production from the natural forests of Guatemala, Mexico, and Honduras meets a significant amount of international demand and supports large numbers of rural families.

The harvesting of *xate, chicle,* and allspice, and the preparation, planting and harvesting of *milpa* (extensive slash and burn agriculture) is

FIGURE 13-2 Blocks or *marquetas* of coagulated *chicle,* the latex of the *chico zapote* (*Manilkara achras*) tree. The *chicle* harvesters (chicleros) carve their initials into the blocks to ensure proper payment.

complementary and overlapping throughout the year (Figure 13-3). While *xate* is collected throughout the year, harvesting peaks in December and again in March and April, corresponding to religious holidays in the United States and Europe. Allspice harvesting starts in late June and extends into September in some areas. *Chicle* camps usually get established in early September, and tapping begins in earnest in mid- to late-September. Forest for *milpa* is cleared in January and February, and burned in March and April. Planting occurs in time for the rains, which, on average, begin in June and run through December. A second or "matahambre" crop is often planted in October.

These three NTFPs occur in remarkably high densities. On the other hand, the plant diversity of the northern Petén does not approach that of rain forests in the Amazon or Costa Rica, where diversity of species may reach hundreds of species of forest trees per hectare. Nonetheless, considerable biodiversity exists in the Petén. This relative dearth of biodiversity works in the favor of sustained production of tropical forest resources. Either as a botanical imprint of the Maya, or the result of other factors, one encounters a forest that is high in both density and

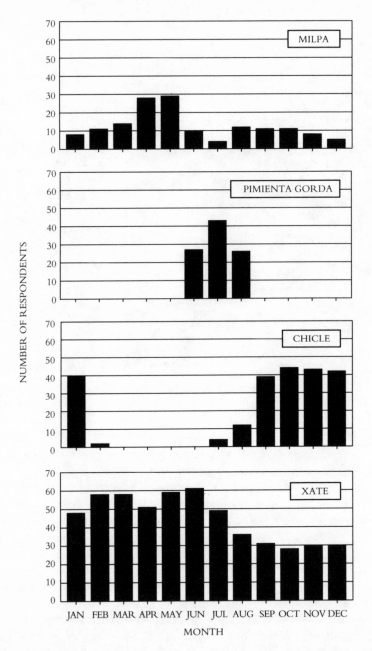

FIGURE 13-3 TIMELINE ILLUSTRATING THE COMPLEMENTARY AND OVERLAPPING NATURE OF MILPA (EXTENSIVE SLASH AND BURN AGRICULTURE) ACTIVITIES AND *XATE*, *CHICLE*, AND ALLSPICE HARVESTING. The left axis indicates how many forest workers in our survey (approximately 90 interviews were conducted) worked in a particular activity in a given month, providing an estimate of the relative importance of each activity over the course of a year.

numbers of marketable species, including *chicle, xate* and *pimienta,* and dozens of "secondary" timbers, thatching palms, construction materials, firewood, and medicinal plants. It is as if this is a premanaged forest. As one exporter of nontimber forest products noted: "Why should I make a plantation? The forests of the Maya Biosphere Reserve are already planted."

The benefits of these products to rural Guatemalans are strong: a family can earn up to three times the average daily wage by harvesting these products versus clearing forest and planting corn or raising cattle. Thousands of families depend on the extractive industries for some or all of their income. To Guatemala as a whole, these renewable resources represent from four to six million dollars each year—and this from products that, if managed properly, harm neither the plant populations nor the larger tropical forest ecosystem. These benefits do not include those from nature and archeological tourism. Tikal National Park, located inside the Maya Biosphere Reserve, draws 15 percent of the all the tourists that come to Guatemala. As a whole, Guatemala earned $185 million from tourism in 1990. Though we cannot say precisely what percentage of this income is due to Tikal, the direct and indirect benefits of nature and archaeological tourism in the Maya Biosphere Reserve are significant. In addition, logging within areas now declared extractive reserves has generated millions each year in foreign exchange, though the benefits are more poorly distributed than with NTFPs. Nevertheless, logging may be possible in extractive reserves, though more research is necessary to learn how to log sustainably without damaging nontimber forest species and the natural forest ecosystem over the long term.

One must consider the relationship between timber and nontimber forest products. A tree is concentrated capital, but you must wait many years to obtain that capital. On the other hand, nontimber forest products can be collected from year to year, though their unit value is relatively small. These two can work together: the nontimber forest products provide rent on the land while one waits for the trees to mature. These mature trees then provide a source of concentrated capital that NTFPs cannot easily provide, while NTFPs make it more economically attractive to leave forest as forest while the trees are maturing.

It is critical to stress that consistent income from harvesting these forest products gives both rural families and the Guatemalan government an economic incentive to conserve the tropical forest of the Petén. Their commitment to do this was reinforced during 1990 when the Guatemalan National Congress passed the Maya Biosphere Reserve law. 750,000 hectares of the Reserve—one half of the whole Reserve—is designated as an extractive reserve for *xate,* allspice, and *chicle* and other important forest products. However, the survival of the area and its harvesting industries is threatened by new colonists moving into the Petén and by would-be developers who do not understand the economic and human benefits of sustained-yield forest harvesting. The forest harvesting industry is also threatened by declining real wages for many forest harvesters. With recent devaluations of the quetzal, local values of *xate* and *pimienta* in quetzales have remained the same while real international prices have stayed the same or risen. This means that the exporters have profited handsomely, while the real net benefits of forest harvesters have dropped. This is not the case with *chicle,* because forest workers, contractors, and the government all receive fixed percentages of the world market value in dollars.

The Maya Biosphere Reserve and its extractive industries face threats from several fronts. Yet, the future is quite hopeful. A new USAID/ Guatemala project called MAYAREMA will offer financial and technical assistance to CONAP in an effort to more sustainably manage the resources of the Reserve. Of critical importance are applied research, extension and training, development of markets for sustainably harvested forest products, promotion of low-impact tourism, and policy changes that will allow local communities and forest harvesters to receive greater benefits from the diverse wildland resources of the Reserve while conserving these resources over the long term. Conservation International, as a primary contractor in this project, hopes to aid Guatemalan decision makers and local communities in their effort to develop the arguments and skills that will conserve the extractive industries and the tropical forest that sustains a whole way of life.

References

Centro Agronomico Tropical de Investigación y Enseñanza (CATIE). 1990a. *Proyecto de conservación para el desarollo sostenible en America Central.* Programa de manejo integrado de recursos naturales. Taller Centroamericano de conservación para el desarrollo sostenible. June 23–26, 1990. Petén, Guatemala.
CATIE. 1990b. *Principales programas para el manejo del parque nacional "Biotopo El Zotz" (San Miguel/La Palotada), Petén, Guatemala.* XIII Curso Anual de planificación y manejo de areas protegidas. July 25-August 29, 1990. Guatemala.
Consejo Nacional de Areas Protegidas (CONAP). 1989. *Estudio Técnico: La Reserva de la Biósfera Maya.* Guatemala: CONAP.
Heinzman, Robert, and Conrad Reining. 1990. *Sustained Rural Development: Extractive Forest Reserves in the Northern Petén of Guatemala.* Tropical Resources Institute Working Paper No. 37. Yale School of Forestry and Environmental Studies. USAID/Guatemala Contract #520–0000–0–00–8532–00.

Nations, James D., and Daniel I. Komer. 1984. *Conservation in Guatemala.* Washington, D.C.: World Wildlife Fund.

Reining, Conrad, Robert Heinzman, Mario Cabrera M., Salvador López, and Axel Solórzano (1991). *A Socio-Economic and Ecological Analysis of Nontimber Forest Products in the Petén, Guatemala.* Washington, D.C.: Conservation International.

Schwartz, Norman. 1990. *Forest Society: A Social History of the Petén, Guatemala.* Philadelphia: University of Pennsylvania Press.

14

The Sustainable Management of Nontimber Rain Forest Products in the Si-a-Paz Peace Park, Nicaragua

JAN SALICK
Department of Botany, Ohio University

The International Peace Park or Si-a-Paz uniting Nicaragua and Costa Rica is one of the largest remaining tracts of tropical rain forest in Central America, with over 500,000 hectares (Morles and Cifuentes, 1989). War was its savior. During the period when Oliver North and his henchmen were running guns and counterrevolutionaries across the Nicaragua–Costa Rican border, the government of Nicaragua patrolled the border heavily and moved campesinos out of the war zone. Never has a national reserve been so well guarded.

As a result of this strict protection, many animal and plant populations that often suffer from human exploitation now flourish in the area. Jaguar, peccary, deer, snakes, and birds are common, as well as valuable timber and nontimber forest products (NTFPs).

Now that the war is over, so is the intense protection of Si-a-Paz. Despite the legalization of the reserve in 1989 by the Sandanista government and in 1990 by the UNO, we see patterns of exploitation and degradation of primary natural resources similar to those in the Amazon or other remote areas.

NONTIMBER FOREST PRODUCTS: ALTERNATIVES TO DEFORESTATION?

What realistic alternatives to tropical deforestation are available? A recent review of this problem suggests natural forest management, extractive reserves, agroforestry, forest restoration, and nature tourism and education are the answers (Gradwohl and Greenberg, 1988; Anderson, 1990). I concentrate on the use and commercialization of nontimber forest products, along with the above mentioned alternatives. Specifically, I compare populations and management of NTFPs within natural forest management for timber, campesino forest management, and unmanaged successional and primary forests. The data are preliminary at this point and I will discuss only trends without drawing final conclusions. Nevertheless, potentials for integrating NTFPs in natural forest management, extractive reserves, and agroforestry are evident.

METHODS

In this ongoing research, systems of natural forest management are being compared by an international research team. The two systems are varying extraction (controlled felling and extraction) and post-harvest silviculture (modified liberation and uniform treatment). Both forest damage and regeneration of NTFPs occur within these experiments. Stratified random sub-subplots (5 meters by 2 meters) totalling 10 percent of each of five permanent subplots for measuring regeneration (0.1 hectares each) are located within each repetition (3 one-hectare plots with 30-meter borders) of each treatment. Within each subplot all the useful plants are inventoried and vouchered with the help of a local, knowledgable informant and data is taken on use and plant community characteristics (species richness, diversity, density, and cover).

For comparison, I am studying the forest management techniques of campesinos within the region. Transects are laid through the campesino forest plots and stratified, and random subplots (5 meters by 2 meters) are located using a random numbers table. All subplots are inventoried, plants vouchered, and data taken exactly as described above. Here, I will report on only an indigenously managed secondary forest of an herbal healer (curandero) near El Castillo, Rio San Juan. Dating back more than four centuries, El Castillo is the site of one of the oldest Spanish forts in Central America (Perez Valle, 1977; MIDINDRA, n.d.). As a result of this historical settlement, El Castillo has long been deforested. Remarkably, the curandero has actively managed his small (approximately 10 hectares) remaining tract of secondary forest for nearly half a century. From this last stand he extracts the medicines he uses, as well as innumerable foods, fibers, construction materials, resins, wood, and so forth.

PRELIMINARY RESULTS

Preliminary results from my ongoing investigations are used to project plant species–area curves for different rain forest successional stages (Figure 14-1). The species–area curves differentiate primary tropical rain forest comprising the most nontimber forest products, followed by the managed forest at El Castillo, with fewer species found in secondary forest following logging (from which the El Castillo forest was derived), and fewest products in early successional forest after agriculture. Other plant community characteristics are compared among primary, secondary, and indigenously managed forests (Figure 14-2). In individual sample plots of 10 square meters, the number of species per plot was equivalent among forest types, although species differed and total species were most numerous in primary forest. Density was somewhat lower in primary forest because of the dominance of large trees, but cover was highest in these forests because of the continuous canopy provided by these dominants. Types or categories of useful plants[1] were remarkably high in all forest, but extremely so in the managed forest. In ethnoecological interviews, the informant was queried as to his techniques.

The curandero of El Castillo employs diverse techniques of forest

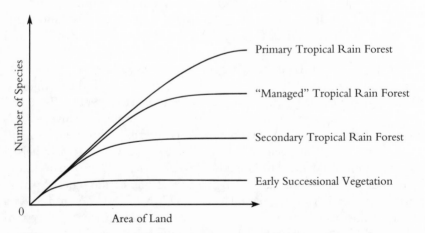

Figure 14-1 Comparisons of Nontimber Forest Products in Different Tropical Forest Types. Useful plant species–area curve trends from four forest types including, in order of greatest to least diversity, primary tropical rain forest, "managed" tropical rain forest of El Castillo, secondary tropical rain forest, and early successional vegetation.

	Species Diversity	Total Species	Density	Cover	Uses
Primary Tropical Rain Forest	=	++	–	+	+
Secondary Tropical Rain Forest	=	–	+	=	+
"Managed" Tropical Rain Forest	=	+	+	=	++

FIGURE 14-2 COMPARISONS OF NONTIMBER FOREST PRODUCTS IN DIFFERENT FOREST TYPES. Plant community trends for useful plant species in three types of tropical rain forests—primary, secondary (selectively harvested), and "managed" forest of El Castillo. Use categories are most abundant in the managed forest, while total species richness is highest in primary forest. Species diversity per 10-square meter sub-subplot is equivalent, although species differ and species diversity also differs over larger areas.

management that he has learned from his Miskito mother and curandero masters or has developed on his own. First he uses indicator plant species to choose a rich and productive tract of forest. He compares his tract to only a few others he knows on the Rios Sabalos and Santa Cruz. Second, he searches far and wide for plants he needs or wants to try and in-plants or transplants these microsites as similar to their original habitat as possible. More in-planting is done with seeds and plants exchanged among curanderos and gathered on travels as far as Panama. Third, he weeds, cultivates the soil, and mulches desirable plants, usually casually, reaching over with his machete to turn the soil as we talk, or pulling weeds as he rests from a strenuous climb. He selects trees to cut by considering the potential regeneration below and liberated desirable saplings. Noxious vines are cleaned after cutting. Fourth, in a small area accidentally burned by fire spread from a neighboring field, he practices more intensive agroforestry by planting beds of ipecac (*Cephaelis ipecacuanha*) below the forest overstory.

DISCUSSION

Data from the locally managed secondary forest in El Castillo, its elevated species area curve, and diverse types of plant uses provide a mark toward which we can aim in developing natural forest management. Can we do as well in maintaining biodiversity and augmenting use as the local curandero? What can we learn from the local forest manager?

One of the goals of buffer zone management is to integrate local people (Gradwohl and Greenberg, 1988). How better than by integrating their own techniques, taking cues from their indigenous forester first (Chambers et al., 1989).

There are additional leads to be taken from the intensive ipecac production by the curandero and on other farms in the region. Sold to the international market, ipecac is a popular nontimber forest product both extracted from natural forest and grown under a range of management techniques bridging forestry and agroforestry. There is a long history of international trade in ipecac, which depends on world market prices, market structure, local economics, the embargo, supply, and other factors. Ipecac, firewood, and mimbre (wicker) represent the three most important commercial NTFPs of the region. Others, used domestically and sold, include thatch, hilera, vines, wood used and sold for other than timber (for example, musical instruments), medicinals (some of which are being tested by the National Cancer Institute [M. J. Balick, pers. comm.]), germ plasm (for example, a wonderfully fragrant wild pineapple) and—more generally—habitat for animals hunted for food and pelts. Local use, management, and commercialization of this array of NTFPs can provide a wealth of information and development strategies to be incorporated with forest management for profit.

The importance of commercial management is not only monetary. The support that NTFPs can provide for other activities, including silvicultural treatments and extractive conservation, is also important. It is often assumed that silviculture or reserve protection is too expensive to be feasible. However, if costs can be defrayed by the profit generated locally by NTFPs or if NTFPs can be used as incentives, both silviculture and protection, as well as natural forest management and conservation, may be realistic options.

BEYOND THE BOTANY

I wish that I could end on a hopeful note, but both the curandero and my own experience in Si-a-Paz tell cautionary tales. What war did not do to Si-a-Paz is being done by economic chaos and planned resettlement of contras and refugees in the buffer zones of the reserve.

As I worked with the curandero of El Castillo, his son, discharged from the drastically reduced army, unemployed, and waiting for his first harvest of newly planted crops, began cutting the long managed forest of its timber to sell. The curandero explained: "We are poor people.

These are hard times, and we must eat." Such is the short-term reality opposing the long-term aspirations of natural resource management.

It is not only the timber that is being rapaciously cut for the desperately needed cash, but also NTFPs. It is not only poor campesinos who are extracting resources at a nonsustainable rate, but the impoverished Nicaraguan forestry corporations as well. It is not only national pressures that Nicaragua experiences, but also the demand of the wealthy timber merchants of Costa Rica who have already exploited the tropical rain forests of their own country and who are now turning to their neighboring countries for illegal harvests. Other timber importing countries are also interested in the harvest of Nicaragua's remaining natural wealth.

Si-a-Paz is one of the last frontiers in long-settled Central America. Logging roads are opening large areas in the buffer zones of Si-a-Paz. After logging by the government, forestry corporation settlers are entering to practice nonsustainable slash and burn agriculture. In the zone, as elsewhere, cattle is often a final chapter in this familiar story, raised less for profit than for investment. Real goods are the only investment available to campesinos who experience 13,000 percent inflation in Nicaragua each year.

In addition to these economic pressures, government and United Nations sponsored repatriation programs for contras and refugees have chosen Si-a-Paz buffer zones and even, in some cases, the reserve itself as ideal sites for thousands of new colonists. The scheme harks back to other rain forest colonization schemes from the Amazon to Borneo, most of which have failed to establish stable, nonindigenous populations in tropical rain forests and have left in their wake deforestation and degradation of natural resources and biodiversity.

These are unquestionably complex and highly charged socio-political issues. We ecologists and botanists studying tropical forestry and nontimber forest products are not eager to enter such discussions. However, an honest evaluation of factors causing deforestation must accompany recommended alternatives to deforestation if our recommendations are not to be simple-minded failures. The present study continues, now incorporating socio-economic analyses along with my own ethno-ecological studies.

Notes

1. Use categories include timber/construction wood, firewood, wood for other uses, construction materials (nonwood), utilities, live-plant utilities, edibles,

intoxicants, medicinals, poisons, gums/resins, oils, edible animal food magics, aesthetics, and others.

References

Anderson, A. (ed.) 1990. *Alternatives to Deforestation: Steps toward Sustainable Use of the Amazon Rain Forest.* New York: Columbia University Press.

Chambers, R., A. Pacey, and L. A. Thrupp. 1989. *Farmer First: Innovation and Agricultural Research.* London: Intermediate Technology Pub.

Gradwohl, J., and R. Greenberg. 1988. *Saving the Tropical Forests.* Washington, D.C.: Island Press

MIDINDRA. (n.d.). *Historia Economica de Rio San Juan.* Midindra, Nicaragua.

Morales, R., and M. Cifuentes (eds.). 1989. *Sistema Regional de Areas Silvestres Protegidas de America Central: Plan de Accion 1989–2000.* CATIE, Turrialba, Costa Rica.

Perez Valle, E. 1977. *El Desaguardero de la Mar Dulce.* Banco de America, Serie Fuentes Historicas No. 7, Nicaragua.

UCA/CATIE/SAREC. 1991. *Plan Operativo para el Desarrollo de Sistemas de Manejo Sostenible para el Aprovechamiento de los Bosques Humedos Tropicales de Nicaragua.* CATIE, Turrialba, Costa Rica.

15

New Nontimber Forest Products from Western South America

ALWYN GENTRY
Missouri Botanical Garden

This chapter will focus on the floristic-taxonomic context in which our knowledge of prospective new useful species of the rain forest is embedded. I will not only point out some potentially useful new nontimber products, but also emphasize how many of these are new, undescribed species. Thus completion of the still woefully inadequate basic floristic inventory of tropical forests becomes an important adjunct to assessing the economic potential of their biodiversity. We cannot use plants if we do not know what they are.

While this point has been made previously, it assumes special significance in the context of northwestern South America. Somewhere between 90,000 and 100,000 plant species are known from the Neotropics, three times as many as in either tropical Africa or tropical Asia (Gentry, 1982; Toledo, 1985), and a disproportionate part of that diversity is concentrated in the northern Andean countries, the region called by Dr. Russell Mittermeier of Conservation International, "the Ilepicenter of biodiversity" (Mittermeier, 1991).

This floristically rich region also has an unusual concentration of locally endemic species, many of which have yet to be officially "discovered." The relatively small northern Andean phytogeographic region of Gentry (1982), which includes the Choco and coastal Ecuador plus the Andes of Colombia and Ecuador above 500 meters elevation, includes about 18 percent of the entire neotropical flora, with 56 percent of the region's described species endemic (Gentry, 1991a).

In addition, there are many more undescribed species. Every major botanical expedition in this region finds species new to science. Even student field trips stumble across new species. For example, during a field course I taught for the National University of Colombia, one day at the La Planada field station we collected all the fertile *Anthurium* species we encountered. Of the 14 *Anthurium* species we found during the course of this class project, 12 proved new to science. Extrapolating from such studies, I estimate that there may be as many as 10,000 undescribed species of vascular plants, the majority of them in northwestern South America (Gentry, 1982; 1986a; 1991a).

Rio Palenque Biological Station in western Ecuador provides a more quantitative example of how little we know. Most of the area is now covered by cow pasture and oil palm plantations. Indeed it took only one decade from when the first road penetrated virgin rain forest in this region for the original forest cover to be almost completely wiped out. Today only the one-square kilometer Rio Palenque Biological Station remains. In the Flora of Rio Palenque, we cataloged 1,100 species of flowering plants from this forest, about 100 of which were new to science (Dodson and Gentry, 1978; 1991).

If we know so little about the plants of western South America, perhaps it is not surprising that even many useful species are undescribed. A good example is *Persea theobromifolia,* formerly the most important timber tree of the Santo Domingo de los Colorados area of western Ecuador. The mahogany-like quality of the wood led to the common name "caoball," and the supposition that the tree was in fact mahogany. However, it turned out to belong to a quite unrelated family, the *Lauraceae.* When I described it as a new species I thought it represented a new subgenus of *Persea,* the avocado genus. Additional study revealed that it could be referred to the segregate genus *Caryodaphnoysis,* previously known only from southeast Asia. Subsequently eight additional neotropical species of that supposedly Asian genus have been discovered, all new to science (Gentry, 1991).

Similarly, the most important construction timber of the Iquitos area, locally called "aceite caspill," turns out to be a species new to science. During the past several years 72 to 74 percent of the wood used for general construction in the Iquitos area came from this species (Vasquez, 1991). "Boa caspill," the main construction timber at Jenaro Herrera on the Rio Ucayali, turns out to be an undescribed species of *Havloclathra,* so distinctive we first suspected it to represent a new genus; it is also the first record of *Havloclathra* for Peru or upper Amazonia (Vasquez, 1991).

Examples like this demonstrate the urgent need for more basic field research, so we can catalog and learn something of the basic biology of the species we hope to learn to use and manage.

There is also a pressing need for more data in planning extractive reserves. In the face of the present taxonomic limitations, the tendency has been to focus plans for sustainable forest utilization, currently a very hot topic in conservation circles, on extractive reserves (Fernside, 1989) featuring few species, for example, rubber and Brazil nuts in Acre, Brazil (Schwartzman, 1989) and *Euterpe* palm near the mouth of the Amazon (Anderson, 1988). The reason that the extractive reserve in the Petén of Guatemala (see Reining and Heinzman, Chapter 13) works so well is that there is relatively low species diversity and a high density of commercial species. Clearly a different approach needs to be adopted if we are to sustainably manage more diverse forest stands.

Another approach achieves sustainability by treating all the species of the forest as commercially equivalent. One example is the Pichis-Palcazu project in Peru (Hartshorn, 1989; 1990). This is a well-planned, ecologically-based model of sustainable yield that attempts to harvest timber in a manner that mimics the ecology of natural gaps in the forest. The fragile soils are also protected by using oxen to haul the timber instead of tractors, with current plans calling for a 30- or 40-year rotation. The problem with this project in terms of making the most of biodiversity is that, in a sense, it reduces the biodiversity of the tropical forest to a kind of the lowest common denominator. Small trees (and branches) are used for charcoal, medium-sized trees for posts and telephone poles, and large trees for timber. In part, the trees are also sorted by species. Thus in essence the diverse forest is treated as though it consisted of only three species. In general the potentially higher value of all the individual species is lost.

At a larger scale the Pulpapel subsidiary of Carton de Colombia is converting much of the lowland rain forest of the Bajo Calima region of Valle, Colombia, to paper pulp for cardboard. Although they are clear-cutting, the company is going to great lengths not to harm the soil and to leave the bark and its nutrients behind. The company anticipates that its approach is sustainable, since much of the biomass of the cut trees, and especially the relatively nutrient-rich bark, is left in place (Ladrach, 1985; Faber-Langendoen, 1989; Faber-Langendoen and Gentry, 1991). However, even if sustainable, such approaches automatically lose the intrinsically higher values of individual species.

The situation of inadequate knowledge is little better for nontimber forest products. Many edible fruits are produced by previously undescribed species. An example with which I was involved is *Passiflora caudata* A. Gentry, a species of coastal Ecuador and southernmost Colombia, close to but definitely distinct from *P. maliformis*, which produces an edible fruit (Gentry, 1980). Another is a new Peruvian species of Hippocrateaceae, recently described as *Salacia alwynii* Mennega, which has the largest flowers in the entire family, twice as large as any previously known. While its fruits are still unknown, despite repeated visits to the type plant, it is likely that they will prove edible. A related species that we recently found on the Rio Madidi of Bolivia proved to have fruits so delicious that several Bolivian biologists took them home to try to grow them as a potential new fruit crop.

NEW VALUES FOR BIODIVERSITY

The people living in the forest typically know that many species have valuable and specific uses (Boom, 1990; Prance et al., 1977; 1987). For example, when the campesinos near Iquitos build a house, they use one tree species for uprights, another for long cross pieces, another for short cross pieces, yet another for thatch. The key point is that by using the individual values of as many different species as possible we can obtain a greater total value.

Perhaps what is really needed is a way to bring more tropical products into the marketplace. Already fruits and other products of different tropical plant species are widely traded in the world economy. Perhaps a mechanism can be found to utilize even more of this biodiversity. Rather than development schemes reducing the diversity of the rain forest to its lowest common denominator, might extractive reserves be designed to achieve full value for the cornucopia of potentially useful tropical forest plants? If so, the extraordinary biodiversity of western South America is a critical resource for the future (Gentry and Blaney, 1990; Gentry, 1991b).

BIODIVERSITY IN WESTERN SOUTH AMERICA

In this context it is noteworthy that western South America is the heart of biodiversity. This is true both at the local and at the regional levels. For example, the most species-rich forest in the world for trees is the Explorama Reserve at Yanamono, Peru, where no fewer than 300 species are included in the 606 plants at more than 10 centimeters diameter in a 1-hectare patch of forest (Gentry, 1988a). Indeed the world's highest local species richness for most kinds of organisms is somewhere in upper Amazonia. For birds and butterflies it is at Tambopata, Peru; for reptiles at Iquitos, Peru; for amphibians at Santa Cecilia, Ecuador (Gentry, 1988a; Lamas, 1985; Dixon and Soini, 1975; 1976; Duellman, 1978); for mammals at Balta on the Peru-Brazil border (Gardner, pers.

comm.); for ants at Cuzco Amazonica near Puerto Maldonado, Madre de Dios (J. Tobin, pers. comm.).

Nor is this phenomenon limited to upper Amazonia. For plants more than 2.5 centimeters in 0.1-hectare samples, the world's greatest species richness is in the Choco area of western Colombia, with the world record from Bajo Calima, where 265 species were included in the 497 individuals in a 0.1-hectare sample (Gentry, 1986b). In part this high level of diversity is the result of unusual and little-understood ecological and evolutionary processes that seem to characterize parts of the Andean fringe region. For example, the Bajo Calima forest is characterized by unusually large leaves, no fewer than a dozen genera or families having their largest known leaves in the world in species endemic to this high rainfall area (Gentry, 1986b). In each case these leaves are similarly large and sclerophyllous, mostly with tannish pubescence on their undersides, a striking example of convergent evolution.

On the other hand, many Andean fringe cloud forest species may be locally endemic due to evolutionary caprice rather than shared selective pressures, with genetic drift and founder effect playing prominent roles (Gentry, 1989). A likely example is the existence of six species of *Gasteranthus* endemic to the Centinela ridge in western Ecuador (one-quarter of the world's total species of the genus) (Gentry and Dodson, 1987). We suspect that evolution may be so labile in this milieu that speciation can occur in nature in as little as 15 years (Gentry and Dodson, 1987).

As a direct or indirect result of the unusually complex mosaic of habitat patches, vegetation zones, and substrates that characterize the northern Andean countries, they have accumulated exceptional biodiversity. Colombia, for example, has more bird species than any other country in the world and is rivalled only by much larger Brazil for plant diversity. Tiny Ecuador may have more orchid species (3,000) than any other country in the world (Dodson, pers. comm.). One way to assess the value and demand for local species is to visit local markets. Not surprisingly, there is biodiversity in the local market. The region's exceptionally high biodiversity is reflected in its plethora of useful plants. Iquitos, Peru might well be termed the useful plant capital of the world. No fewer than 193 species of fruits are consumed at Iquitos, 139 of them native, and 57 wild-harvested species from 24 different families are sold on the local market (Vasquez and Gentry, 1989). Fruits of no fewer than 39 plant families are eaten, 34 of which include species with wild-harvested fruits, an impressive taxonomic diversity. Half of the numerous fruit vendors in the Iquitos market sell wild-harvested fruits and we estimate that at least one-tenth of the diet of the average Iquitos-area campesino comes from wild-harvested fruits (Vasquez and Gentry, 1989).

Especially important are palms like *Mauritia,* whose fruits are a mainstay of the diet of many Loretanos and that also produces a delicious ice cream. Some native fruit species like Couma *macrocarpa, Pourouma cecropiifolia,* and *Grias neuberthii* have begun to be cultivated, and several native Amazonian fruits that do not grow wild in the Iquitos vicinity are also locally cultivated.

Most of these fruits are little known outside the region and some of them represent species new to science. Two well-known *IrVanthera* species, members of a genus not generally thought to produce edible fruits, have fruits eaten at Iquitos. One of these is also reported by Boom (1990) as an important fruit in Amazonian Bolivia. Might some of the rarer or newly described species of the genus also have edible fruits? Another Myristicaceae, *Compsoneura atopa* A. C. Smith, endemic to the Colombian Choco, has a large edible seed that tastes potato-like when cooked (Cuadros, pers. comm.); a closely related species, probably undescribed, has a similarly edible seed.

An interesting feature of the trans-Andean region is that it seems unusually rich in fruits with starchy carbohydrate-rich seeds, as opposed to the sugary pulp of most fruits. Perhaps this is correlated with the general tendency to large fruits that characterizes the region's plants (Gentry, 1986b). In addition to *Compsoneura,* good examples come from the Bombacaceae, where edible-seeded Patinoa represents an entire genus described in 1953 for an edible-fruited Choco plant, although several other species were subsequently discovered in Amazonia (Cuatrecasas, 1953; Romero-Castaneda, 1985). *Pochote* (formerly *Bombacopsis*) *patinoi* Duqand and Robyns is a similar case (O. de Benavides, pers. comm.). The endemic Choco species *Orbigynya cuatrecasana* has an edible coconut-like fruit that may be the largest of any palm other than the coconut and the double coconut.

Another interesting example is *Erythrina edulis.* It is well known in the Andes because it is such a good source of protein, and several projects are currently underway with this legume to expand the area in which it is cultivated and consumed. The point I wish to make is not so much that *Erythrina edulis* is good to eat, but that it has numerous relatives that may also be good to eat but have yet to be investigated—some of them newly discovered like *Erythrina megistophylla* from Rio Palenque or the unnamed yellow-flowered species I found on a recent visit to Maquipucuna, another western Ecuador field station.

To what extent is high biodiversity reflected in a high diversity of useful plants? This is a key question that is very difficult to address given our abysmal lack of knowledge of the plants of western South America, much less their potential uses. However, a recent summary of Biqnoniaceae ethnobotany provides a tantalizing hint (Gentry, 1992). There are about 200 Bignoniaceae species known from the northern Andean countries. Excluding horticulture, there are reports of specific uses for at least 75 of these Bignoniaceae species, although many of the reports come from elsewhere in the range of a given species. Thus three-eighths or 38 percent of the Bignoniaceae of northwestern South America are documented as useful, including many rare taxa and at least six species I have only recently described. We may assume that many more uses remain to be discovered. Thus approaching half of the region's Bignoniaceae are reported to have specific uses. If this figure is generally applicable, then half of the region's species might be expected to prove directly useful to man, a figure that projects to perhaps 10,000 or more useful species in this region alone. Obviously the anticipated uses of most of these species remain to be discovered, just as do many of the species themselves.

ECONOMIC VALUE OF BIODIVERSITY

There is an urgent need for determining the actual and potential economic value of rain forest species. One of the most promising of these species is "camu-camu" (Myrciaria dubia of the Myrtaceae) that thrives in seasonally inundated forests in Amazonia and has fruits containing over ten times the vitamin C of citrus (Peters and Hammond, 1990). Dr. Charles Peters of the New York Botanical Garden has concluded that a 1-hectare stand of camu-camu produces between $5,700 and $7,620 worth of fruit per year with a 60-hectare study site actually producing $10,000 of camu-camu in a single year, and in an area of little use for intensive agriculture (Peters, Balick et al. 1989; Peters and Hammond, 1990).

Peters and I also collaborated on a paper (1989) in which we attempted to document the actual value of a single hectare of Amazonian forest. The forest, located near Mishana on the Rio Nanay near Iquitos, is highly diverse, with 289 species among its 858 individual trees and lianas more than 10 centimeters dbh. (Gentry, 1988a), 454 of the trees having actual or potential uses (Gentry, 1986c). We calculated the annual economic yield from the 72 (26 percent) tree species and 350 individuals (42 percent) that produce products sold in the Iquitos market. Eleven species produce edible fruits, 60 species commercial timber, and Hevea *quianensis* rubber. As a whole, the hectare of forest produces $650 worth of fruit and $50 of rubber per year. There are also 93 cubic meters of sellable timber worth $1,000. We later added an economist as a third author. We concluded that the net present value (NPV) of the hectare of species-rich forest is $9,000; allowing 25 percent of the fruit to remain for future regeneration, this comes to a very conservative $6,330 NPV. Therefore, the value of the sustainably utilized "unimproved" forest is several times greater than the value of the same land converted into cattle-ranching (NPV = $2,960) or managed plantations (NPV = $3,184) (Peters et al., 1989). Since the values for cattle-ranching and managed plantations would be far less for a white-sand area like Mishana, it is probable that at Mishana the land with intact forest is worth several hundred times more than under any alternative usage.

Moreover, we probably underestimated the value of the standing forest because we did not include the value of species utilized for fiber and/ or medicine, which are mostly harvested from plants less than 10 centimeters dbh. For example, the main local industry at Mishana is harvest of the leaves of "irapay" (*Lepidocaryum tessmannii*) for thatch for the roofs of Iquitos, but this value was omitted from our calculations since this palm is too small to be included in our census data. Many other smaller plants are also commercially valuable. Edible fruits sold in the Iquitos market are produced in our plot by such stemless palms as *Scheelea plowmanii* Glassman, yet another species that has not yet been formally "described." Many of the lianas in our hectare plot (for example, *Strychnos, Martinella,* Menispermaceae) are rich in biologically active alkaloids, some of which are commercialized elsewhere.

Several climbers and hemiepiphytes that are too small for inclusion in our data yield locally important fibers. While the value of fibers might not be obvious, Myers (1984) suggests that rattan products are southeast Asia's most valuable export after timber. In the Neotropics, a completely unrelated climbing palm, *Desmoncus,* has similar properties, and its fibers, including those of the recently *D. cirrhiferus* Gentry and Zardini, are used in basket-making by indigenous groups (Gentry, 1988b). Recently, a cottage industry in rattan products has sprung up around Iquitos, Peru, where there is already a thriving fiber industry based on the aerial roots of *Philodendron solimoesensis* A. C. Smith (Araceae), which have long been extensively used in campesino handicrafts. The stems of various other monocots, including *Ischnosiphon* (Marantaceae) and *Heteropsis* (Araceae), are similarly used locally. It seems possible that export of such handicraft items could be one element of a strategy for sustained use of the intact forest. Many of these fiber plants occur naturally in second-growth forest, an increasingly prevalent habitat, whose utilization could help deflect some of the development pressure from primary forests.

Finally, it is worth noting that tropical forests are the world's most productive system for converting the sun's energy into the hydrocarbons on which humankind is so dependent (Rosenzweig, 1968). Perhaps energy production could be a part of a sustainable system for utilizing tropical forests. For example, a genus of Cucurbitaceae climbers (*Fevillea*) has seeds richer in oil than any other dicot, so rich that the Campas Indians of Amazonian Peru burn them as candles. We (Gentry and Wettach, 1986) have calculated that if *Fevillea* were to replace the lianas in an otherwise natural forest and continue to produce oil at its normal levels, such a forest would produce as much oil per hectare as any extant oil plantation, saving the forest trees in the bargain. The dwindling oil and coal resources to which so much attention is given actually constitute energy fixed by ancient tropical forests. It seems strange that so little attention has been given to this facet of tropical forests. There seems no intrinsic reason why we should leave tropical forest–fixed hydrocarbons underground for a few million years before we use them.

The final point that I wish to make is that there is an important need for basic floristic inventory if new tropical forest products are to be included in schemes for sustainable use of tropical forests. Local people seldom know *everything* there is to know about the potential uses of local plants. For example, it is noteworthy that the wild-harvested fruits eaten at Iquitos are different from those consumed elsewhere in Amazonia. Cavalcante (1972, 1974, 1979) reports that almost exactly the same number of fruit species (167) and families (37) are consumed in Amazonian Brazil as around Iquitos. However, only a quarter of the 121 wild-harvested fruits listed by Cavalcante are the same as those used around Iquitos, despite the fact that at least 15 additional species listed by Cavalcante grow wild around Iquitos but are not locally consumed.

A good example of this problem is *Melloa quadrivalvis* (Jacq.) A. Gentry, a Bignoniaceae liana that ranges from Mexico to Argentina but is apparently used in only one place. In northern Colombia near Cartagena, several families of local fishermen leave a powder of this plant outside the burrows of sand crabs. The crabs eat it and are temporarily paralyzed, waiting for the fishermen to collect them the next morning. The potion seems perfectly biodegradable since the crabs recover by the time they arrive at the local market, and we ate crabs harvested in this manner, suffering no ill effects to date.

The same pattern of differential knowledge of plant uses holds true throughout Latin America. For example, *Brosimum alicastrum* ("ramon") probably provided one of the staples of Mayan civilization (Puleston, 1982; Gomez-Pompa, 1987). Although this and other *Brosimum* species are common components of Amazonian forests, they are not locally consumed and thus not included in our tabulation of useful fruits nor the valuation of the Mishana hectare. Similarly, *Dialium guianense*

(Aubl.) Sandw. is valued in Mexico as the source of an edible fruit. Although it was in our Mishana tree plot, it is not used in the Iquitos region and thus was not included in our tabulations.

We may conclude not only that there are a wealth of useful species in the forests of western South America, but also that many of the potentially useful species of an area may be locally unknown. If we seek to use new species, it may not be enough to find out how the local people use plants, although that is an important first step. Equally important are such esoteric approaches as comparative floristic inventories that can serve to catalog which plants of which region or soil type might have the most potential as new resources. Such studies have already demonstrated, for example, that tropical forest plants are highly substrate-specific and thus appropriate for cultivation or development only on a small subset of soil types. The diversity of tropical forests need not be an obstacle to their sustainable utilization if we dedicate ourselves to cataloging it, understanding it, and learning how it might be used.

References

Anderson, A. 1988. Use and Management of Native Forests Dominated by Acai Palm (*Euterpe oleracea* Mart.) in the Amazon Estuary. *Adv. Econ. Bot.* 6:144–54.

Boom, B. 1990. Ethnobotany of the Chacobo Indians, Beni, Bolivia. *Adv. Econ. Bot.* 4:1–68.

Cavalcante, P. B. 1972, 1974, 1979. *Frutas Comestíveis da Amazonia I, II, III.* Publ. Avulsas Mus. Goeldi 17, 27, 33.

Cuatrecasas, J. 1953. Un nouveau genre de Bombacees, *Patinoa. Revue Internat. Bot. Appl. d'Agr. Trop.* 33:306–13.

Dixon, J. R., and P. Soini. 1975. The Reptiles of the Upper Amazon Basin, Iquitos Region, Peru. I. Lizards and Amphisbaenians. *Contr. Biol. Geol. Milwaukee Publ. Mus.* 4:1–58.

Dixon, J. R., and P. Soini. 1976. The Reptiles of the Upper Amazon Basin Region, Peru. II. Crocodilians, Turtles, and Snakes. *Contr. Biol. Geol. Milwaukee Publ. Mus.* 12:1–91.

Dodson, C., and A. Gentry. 1978. Flora of the Rio Palenque Science Center. *Selbyana* 4:1–628.

Dodson, C., and A. Gentry. 1991. Biological Extinction in Western Ecuador. *Ann. Missouri Bot. Gard.* 78:273–95.

Duellman, W. E. 1978. The Biology of an Equatorial Herpetofauna in Amazonian Ecuador. *Univ. Kansas Mus. Nat. Hist. Misc. Publ.* 65:1–352.

Faber-Langendoen, D. 1989. Combining Conservation and Forestry in a Colombian Rain Forest: An Ecological Assessment. Ph.D. Thesis, St. Louis University.

Faber-Langendoen, D., and A. Gentry. 1991. The Structure and Diversity of Rain Forests at Bajo Calima, Choco Region, Western Colombia. *Biotropica* 23:2–11.

Fernside, P. M. 1989. Extractive Reserves in Brazilian Amazonia. *Bioscience* 39:387–93.

Gentry, A. H. 1980. New Species of Apocynaceae, Bignoniaceae, Passifloraceae, and Piperaceae from Coastal Colombia and Ecuador. *Phytologia* 47:97–107.

Gentry, A. H. 1982. Neotropical Floristic Diversity: Phytogeographical Connections between Central and South America, Pleistocene Climatic Fluctuations, or an Accident of the Andean Orogeny? *Ann. Missouri Bot. Gard.* 69:557–93.

Gentry, A. H. 1986a. Endemism in Tropical versus Temperate Plant Communities. In M. Soule (ed.), *Conservation Biology.* Sunderland, MA: Sinauer Press.

Gentry, A. H. 1986b. Species Richness and Floristic Composition of Choco Region Plant Communities. *Caldasia* 15:71–91.

Gentry, A. H. 1986c. Sumario de patrones fitogeograficos neotropicales y sus implicaciones para el desarrollo de la Amazonia. *Rev. Acad. Col. Cienc.* 16:101–16.

Gentry, A. H. 1988a. Tree Species Richness of Upper Amazonian Forests. *Proc. Nat. Acad. Sci.* 85:156–59.

Gentry, A. H. 1988b. New Species and a New Combination for Plants from Trans-Andean South America. *Ann. Missouri Bot. Gard.* 75:1429–39.

Gentry, A. H. 1989. Speciation in Tropical Forests. In L. Holm-Nielsen, I. Nielsen, and H. Balslev (eds.), *Tropical Forests: Botanical Dynamics, Speciation, and Diversity.* London: Academic Press.

Gentry, A. H. 1991a. Tropical Forest Biodiversity: Distributional Patterns and Their Conservational Significance. Oikos 162:000000.

Gentry, A. H. 1991b. Tropical Forest Diversity versus Development: Obstacle or Opportunity? *Diversity* 7:89–90.

Gentry, A. H. 1992. Synopsis of Bignoniaceae Ethnobotany and Economic Botany. *Ann. Missouri Bot. Gard.* 79:000–000.

Gentry, A. H. and C. Blaney. 1990. Alternative to Destruction: Using the Biodiversity of Tropical Forests. *Western Wildl.* 16:2–7.

Gentry, A. H., and C. Dodson. 1987. Diversity and Biogeography of Neotropical Vascular Epiphytes. *Ann. Missouri Bot. Gard.* 74:205–33.

Gentry, A. H., and R. Wettach. 1986. Fevillea—A New Oilseed from Amazonian Peru. *Econ. Bot.* 40:177–85.

Gomez-Pompa, A. 1987. Tropical Deforestation and Maya Silviculture: An Ecological Paradox. *Tulane Studies Zool. and Bot.* 26:19–37.

Hartshorn, G. 1989. Application of Gap Theory to Tropical Forest Management: Natural Regeneration on Strip Clear-Cuts in the Peruvian Amazon. *Ecol.* 70:567–69.

Hartshorn, G. 1990. An Overview of Neotropical Forest Dynamics. In A. Gentry (ed.), *Four Neotropical Rainforests.* New Haven, CT: Yale Univ. Press.

Ladrach, W. E. 1985. History and Management of the Bajo Calima Concession. In W. Ladrach (ed.), *Forest Research in the Bajo Calima Concession,* 9th annual report. Carton de Colombia, S.A., Cali, Colombia.

Lamas, G. 1985. *Rev. Peruana Entom.* 27:59–73.

Myers, N. 1984. *The Primary Source: Tropical Forests and Our Future.* New York, London: Norton.

Peters, C. M., M. J. Balick, F. Kahn, and A. Anderson. 1989. Oligarchic Forests of Economic Plants in Amazonia: Utilization and Conservation of an Important Tropical Resource. *Cons. Biol.* 3:341–49.

Peters, C. M., A. Gentry, and R. Mendelsohn. 1989. Valuation of an Amazonian Rainforest. *Nature* 339: 655–56.

Peters, C. M., and E. Hammond. 1990. Fruits from the Flooded Forests of Peruvian Amazonia: Yield Estimates for Natural Populations of Three Promising Species. *Adv. Econ. Bot.* 8:159–76.

Prance, G., D. Campbell, and B. Nelson. 1977. The Ethnobotany of the Paumari Indians. *Econ. Bot.* 31:129–39.

Prance, G. D., W. Balee, and B. Boom. 1987. Quantitative Ethnobotany and the Case for Conservation in Amazonia. *Cons. Biol.* 1:296–310.

Puleston, D. E. 1982. The Role of Ramon in Maya Subsistence. In K. V. Flannery (ed.), *Maya Subsistence.* New York: Academic Press.

Romero-Castaneda, R. 1985. *Frutas Silvestres del Choco.* Instituto Colombiano de Cultura Hispanica, Bogota.

Rosenzweig, M. L. 1968. Net Primary Productivity of Terrestrial Communities: Prediction from Climatological Data. *Am. Nat.* 102:67–73.

Schwartzman, S. 1989. Extractive Reserves: The Rubber Tappers' Strategy for Sustainable Use of Amazon Rain Forest. In J. O. Browder (ed.), *Fragile Lands of Latin America: Strategies for Sustainable Development.* Boulder, CO: Westview Press.

Toledo, V. 1985. *A Critical Evaluation of the Floristic Knowledge in Latin America and the Caribbean.* Nature Conservancy Intern. Progr. Unpubl. Report, Washington, D.C.

Vasquez, R. 1991. El genero *Caraipa* en Peru. *Ann. Missouri Bot. Gard.* 78.

Vasquez, R., and A. H. Gentry. 1989. Use and Misuse of Forest-Harvested Fruits in the Iquitos Area. *Cons. Biol.* 3:1–12.

16

Native Plant Products from the Arid Neotropical Species: Assessing Benefits to Cultural, Environmental, and Genetic Diversity

GARY PAUL NABHAN
Native Seed/SEARCH

Over the last two decades, numerous desert plants have been developed for their unique culinary, pharmacological, and industrial products in the Sonoran Desert of the United States and Mexico. Some of these, such as jojoba (*Simmondsia chinensis*) have been successfully brought into cultivation on more than 25,000 hectares of desert lands, and numerous shampoos and cosmetics are now internationally marketed. However, few of the initial cultural and environmental conservation objectives have been achieved. To avoid similar situations in which commercialization occurs without cultural, ecological, or genetic diversity benefits, an analysis was undertaken to determine which economic botany projects with arid subtropical plants from the Sonoran Desert were analyzed with regard to whether or not they have been domesticated or wild-harvested only, and what benefits have been derived. The results indicate that few modern developments of desert crops have yet provided new benefits to the cultural and ecological communitites from which they are derived. Carefully planned small-scale industries based on community participation and in situ conservation have the greatest potential for achieving such goals.

Over the last decade, there has been considerable scientific interest in developing new products from economic plants in both the humid and arid Neotropics. Although economic botanists from these two climatic zones have seldom compared objectives, methodologies, and strategies, they share many of the same goals and assumptions. First and foremost is the recognition that there are many under-exploited wild plant resources that have unusual properties and ecological characteristics that could potentially contribute to sustainable economies for their regions. Second, many economic botanists carry the hope that the development of native plants in situ might discourage the conversion of native vegetation for conventional agriculture, and therefore reduce the loss of biodiversity that typically results from such land conversion. By developing new industries based on the neotropical flora, many scientists assume that they can help relieve the poverty endemic to certain Latin American regions, and improve the quality of life both economically and ecologically.

While these lofty goals are admirable, they nevertheless can be considered to be "biotechnological fixes" that offer simple solutions to complex problems. Two routes of native plant development have been traveled, and both have their attendant curves, ruts, and pitfalls. The first route covers intensified extraction of wild plants from natural vegetation, and may or may not include strategies to increase yield efficiencies, reduce over-exploitation, and promote reforestation or rustic agroforestry. In some cases, the new demand for a wild product has had unintended negative effects, including illicit harvesting from parks and reserves, depletion of natural populations of the target plant, and disruption of associated plant and animal species, or local competition for scarce resources.

The second route—that of plant domestication for agronomic production—is not without its own hazards. Even though wild economic plants may seemingly demonstrate tolerance to drought, heat, pathogens, pests, or other stresses in their natural habitats, vulnerability to these stresses may increase dramatically when the plants are grown under intensive, monocultural production.

As for improving the socioeconomic status of local campesinos and indigenous people, some economic plant developments have resulted in diverting benefits away from these communities rather than toward them. The development of the liquid wax of jojoba, *Simmondisia chinensis,* originally promoted for saving whales and for providing stable sources of revenue to Apache Indians, has achieved neither of these goals. Rather than using the wild or cultivated jojoba harvests of the Apache for production of candles, cosmetics, and lubricants, the jojoba industry is now based on venture capital of California, Arizona, and

Japanese upper classes, which have used immature irrigated plantations as tax write-offs. While now grown on more than 25,000 hectares of Sonoran Desert lands, only a small portion of the production is in any way benefiting the indigenous people who have used jojoba as a medicine and cosmetic for hundreds of years.

If one wishes to avoid such pitfalls, what criteria can be used to fit a native plant into a new production system so that some social, ecological, and genetic conservation benefits are assured? The criteria chosen and the ultimate success of the project depend on how clearly the goals of natural resource development are defined *in advance*. Is the plant intended to provide both cash income and direct nutritional benefits to local peoples, or only one of these? Is agroforestry production of the resource aimed at halting the depletion of the wild populations by reducing harvest pressures in their natural habitats? Is sustained plant extraction only one component of a larger conservation and development scheme, or is it the prevailing goal?

Any combinations of these goals have been cited as valid rationales for economic plant development schemes. From my point of view, however, the following criteria offer the best guidelines for ensuring that indigenous peoples and other peasant communities benefit from applied ethnobotanical development, and that projects sustain rather than deplete biodiversity: 1) The project should attempt to improve the objective and subjective well-being of local communites—the latter as members perceive themselves—rather than simply seeking cheap production sites and importing inexpensive labor. 2) Cultivation in fields or through agroforestry management should be considered if there is the threat that wild harvests will deplete the resource in its habitat. At the same time, the new production should offer a new source of employment to those who were formerly dependent on the wild harvest. 3) Wildland management and more sensitive harvesting practices should be introduced if the resource might sustain economic levels of extraction in the habitat, eliminating the need to convert that habitat to conventional agriculture or livestock production. However, the biotic communitiy should be monitored to determine whether the regenerative capacities of populations of the targeted resource or its associated species are measurably affected. 4) The plant(s) chosen should offer multiple products or be adapted to diversified production systems, so that small producers are buffered from the vagaries of any single economic market. 5) Whenever possible, the project should build on local familiarity, use, and conservation traditions for the plant being developed. 6) If possible, the economic botany project should draw on locally available genetic resources, technologies, and social organizations, so that local people can retain control over the destiny of the resource.

By establishing the above-mentioned criteria, we may have a comparative basis for evaluating the success of conservation and development efforts. It is hoped that these checklists can be applied not only to economic plants from the arid Neotropics, but from the humid Neotropics as well.

III

Palms and Their Potential

17

Babassu Palm Product Markets

PETER H. MAY

Non-Wood Forest Products Division, FAO Forestry Department

An adequate market assessment of the potential of nontimber forest products (NTFPs) is important prior to investment in product development. The terms under which such products are traded in the market, and the tenure rights over the resources from which they are derived determine, in large measure, whether such resources may be managed for sustained production. Prospects for trade expansion in NTFPs suffer from the following restrictions: 1) Poor market infrastructure, perishable products, and isolation from information on prices or market trends, which make forest dwellers vulnerable to buyers' manipulations and limit the scope for increased output. 2) Markets may be highly specialized, fickle, and limited in size or scope for expansion, and if output expands, the bottom may drop out of the market, leading to abandonment of production. 3) If markets expand, the rush to exploit new opportunities may exhaust fragile and poorly understood resources that previously supplied low volumes of useful products in a sustainable fashion.

BABASSU PALM—SUBSIDY FROM NATURE

Among the more promising NTFPs for development are those that occur in so-called "oligarchic" forests (Peters, in press), which are dominated by single or few species whose growth and reproductive characteristics protect them from human activity such as slash and burn cultivation. These stands, considerably less fragile than primary tropical broadleaved forests, tend to arise in succession after clearing for cultivation or pasture. Their resilience to environmental and human-induced stresses, location near settlement areas, and products useful for subsistence and market purposes make them particularly interesting for rural development efforts, comprising what has been termed a "subsidy from nature" (Hecht et al., 1988; Anderson et al., 1991).

Among oligarchic forest-forming species, the babassu palm (*Orbignya phalerata*) is particularly promising as a source for a wide range of subsistence products (Table 17-1), and for its importance in the regional and national economy as a source of industrial oil, feed, and fuel. Babassu palms have formed dense stands reaching up to 66 percent dominance in areas of mid-north Brazil, Tocantins, and Mato Grosso, although stands have been characterized as far afield as Rondonia and Bolivia. Because babassu palms produce massive volumes of biomass in the form of leaves, averaging over 17 metric tons per hectare annually (Anderson et al., 1991), shifting cultivators are able to cut leaves rather than stems to produce the intense fire required for nutrient provision to crops. The palms are then allowed to rejuvenate, providing important market and subsistence products during the fallow cycle.

It is in Maranhao and Piaui that babassu palm forests are most impressive in area and economic importance. Although counting only half of all inventoried palm stands (MIC/STI, 1982), Maranhao producers generated nearly three-quarters of all kernels extracted in 1980 (IBGE, 1983). According to surveys carried out in the early 1980s, babassu kernel sales alone provide an average of between 27 percent and 30 percent of cash household income; and babassu products in general contributed 22 percent of income from all sources to as many as 420,000 rural families in Brazil (May, 1986; Queiroz, 1989). Babassu charcoal is increasingly entering into market circuits as Brazil searches for alternative fuel sources to feed a growing siderurgical industry. A wide range of potential byproducts exist that may be derived from babassu fruit through *aproveitamento integral* (full utilization).

Unfortunately, however, a great deal of babassu's potential has not been and may never be realized. This somber assessment is primarily due to the secularly low value of the palm's principal products in relation to forest productivity and fruit bulk. It is also attributable to the character of the forest products industries that have arisen based on babassu, the structure of market intermediation (May, 1989), and the property rights over the palms themselves. The result of these problems has been a decline in the babassu industry, low remuneration to producers, and consequent deforestation of high-density stands to make way for more remunerative land uses. Only development measures that address these problems in concert can be expected to ensure that babassu products may become more viable components of rural enterprises (May, 1990).

TABLE 17-1   *Subsistence Uses of the Babassu Palm*

LEAVES

Fibers

Baskets for storage and transport

Mats for doors, windows, rugs, grain-drying

Fans for ventilating fires

Sieves for sifting manioc flour and rice

Other uses: twine, torches, whisks, bird cages, hunting blinds, animal traps

Construction Materials

Thatch for roofing and walls

Laths for window frames and support of clay-packed walls

Rails for fencing to protect agricultural plots from animals and delimit hunting zones

Agricultural Uses

Leaves burned in shifting cultivation plots to promote nutrient recycling and pest control

Stakes used for crop support and elevation of planters

Living leaves provide shade for livestock and feed during dry periods

Medicine

Liquid expressed from rachis used as antiseptic and styptic

STEMS

Construction

Entire stems used to make bridges, foundations, benches

Food

Palm heart used as food for people and animals

Sap collected from stumps of felled palms used in fermented drink

Burned stems for manufacture of salt by indigenous people

Attraction of game

Sap collected from stumps of felled palms used as attraction of beetle larvae that are eaten or used as fish bait

Horticulture

Decayed stems used as mulch and planting medium

TABLE 17-1   (continued)

FRUITS

Food

Kernels consumed raw as snack nut

Milk produced from kernels used as a beverage or for stewing meat and fish

Oil extracted from kernels used for cooking

Residues of kernels used as animal feed and substitute or filler for coffee

Beetle larvae extracted from kernels used as food

Flour made from mesocarp used as a substitute for manioc

Flour made from mesocarp used to make chocolate-like beverage

Medicine

Liquid endosperm used to treat sties and bleeding

Tar from burning husks rubbed on gums to alleviate toothaches

Flour made from mesocarp used to treat gastrointestinal complaints

Fuel

Charcoal from husks used as principal source of fuel for cooking

Oil extracted from kernels used for burning lamps

Entire fruits used as fuel to smoke rubber

Attraction of Game

Fruit mesocarp serves to attract rodents

Residues of kernels used as shrimp and fish bait

Handicrafts

Fruit endocarp used to make pencil holders, keychains, figuirines

Other Uses

Oil extracted from kernels used to make soap

Beetle larvae extracted from kernels rubbed on bows to increase resiliency

Smoke from charcoal production serves as insect repellent

SOURCE: Adapted from May et al., 1985

TRADITIONAL OIL MARKETS

Babassu kernel output passed through three discrete historical phases. Between 1920 and 1935, kernels were predominantly traded in international markets after being shipped to southern Brazil. From 1940 to 1960, kernels were increasingly used in domestic vegetable oil industries in Rio de Janeiro and São Paulo for edible and cosmetics products. Finally, from 1960 to the present, the babassu kernel oil industry was almost completely transferred to facilities within the babassu zone itself, which included as many as 32 firms operating in Maranhao alone in 1981 (Amaral, 1983), and an additional 20 operating in Ceará, Piaui, Goias (now Tocantins) and other states. There, crude vegetable oil was expressed, and the majority shipped to industries in the south for use in soaps and other cosmetics. The press cake obtained as a by-product (about 32 percent of raw material by weight) was shipped primarily to European markets as feed.

In the 1970s, subsidized capital availability enabled some of these firms to invest in more sophisticated and larger-scale equipment, and to expand vertically into final product lines such as refined oil, soaps, and glycerine. However, the economic crisis of the 1980s, substitution of oil-based soaps by synthetic detergents, substitution of saturated oils such as babassu by soybean oil in edible uses, and land-use change within the babassu zone have resulted in traditional oil markets facing a tendency toward constriction. The result is that the number of oil pressing firms in operation has been reduced considerably (Table 17-2). Finally, recent opening of trade restrictions against importation of lauric oils such as coconut and African palm kernel oil would tend to force babassu oil prices down, making the retention of palm-based enterprises even more tenuous.

CHARCOAL AND OTHER FRUIT PRODUCTS

During the 1970s, domestic demand for biomass fuels accelerated because of the petroleum price shocks, and a number of options were considered, including babassu fruit husk charcoal, tar, and ethanol. Technological advances that were based on over a century of efforts to mechanically break the dense babassu fruit and separate its component parts now make it feasible to disseminate such processes to make more complete use of these products. Despite the promising caloric potential of these products, however, the low productivity of native babassu

TABLE 17-2   *Babassu Oil Production in Metric Tons*

Year	Total Production	Export	Domestic Consumption
1965	54,102	12,017	42,085
1966	66,614	5,475	61,139
1967	51,479	4,198	47,281
1968	65,394	8,986	56,408
1969	100,678	15,497	85,181
1970	87,391	14,419	72,972
1971	72,201	1,450	70,751
1972	80,000	2,072	77,928
1973	86,500	1,470	85,030
1974	118,858	40,282	78,576
1975	85,000	950	84,050
1976	95,000	369	94,631
1977	90,000	4,446	85,554
1978	95,000	9,218	85,782
1979	115,000	20,427	94,573
1980	100,000	2,576	97,424
1981	95,000	600	94,400
1982	80,000	0	80,000
1983	90,000	11,900	78,100
1984	100,000	12,700	87,300
1985	82,000	2,000	80,000
1986	70,000	0	70,000

SOURCE: Frazao and Pinheiro, 1988

stands, transport difficulties for whole fruit, resistance to selling fruit, and problems in penetrating untraditional markets stymied efforts to create vertically integrated industries based on babassu.

One of the most controversial barriers to progress in complete utilization of babassu fruit is that of the social impact that would result from their industrialization, which could result in employment displacement on a massive scale (Mattar, 1979). On the other hand, it is estimated that as much as two-thirds of total babassu fruit output remains unexploited, due to limitations in infrastructure, access to stands, and property rights (Pick et al., 1985). Unless their prices could be substantially reduced through some form of mechanical processing and increased efficiency in marketing, however, it is dubious whether considerably greater volumes of babassu products would find ready buyers.

At the present time, babassu charcoal derived from husks carbonized in small-scale rural kilns by fruit breakers is the only other product of the palm that is consistently marketed in any volume. A new market in activated charcoal is currently being developed by an Italian/Swiss firm based in Timon, Maranhao, with international markets in North America, Australia, and South Africa paying between $1,500 and $2,000 (U.S. dollars) per ton for this product. Based initially on charcoal purchased from breakers, the investors—disappointed with the high level of impurities associated with this raw material—hope to soon incorporate the processes of whole fruit industrialization in their operation.

ALTERNATIVE MARKETING APPROACHES

There may be opportunities for small volumes of oil to be traded in nontraditional circuits at more remunerative prices. In contrast with declining demand and competition elsewhere, babassu oil is highly prefered for edible uses in the palm zone, and is used in local small-scale soap manufacture. Relatively modest demand may also be generated by "green" industries overseas, such as the Body Shop.

A project of the Association of Agrarian Reform Settlements of Maranhao (ASSEMA) in conjunction with Ecotec/IDE and Cultural Survival, among other collaborators, seeks to organize small-scale babassu kernel oil pressing operations at the community level, and to market output cooperatively. In conjunction with the ASSEMA project, a small-scale babassu fruit breaking system appropriate for community enterprises has been developed by the Maranhao State Agricultural Research Corporation (EMAPA) with support from the national funding agency (FINEP). This system produces kernels, endocarp (for high-quality charcoal production), and starchy mesocarp meal with considerable efficiency, using equipment developed by a mechanic in Fortaleza, Ceará, costing on the order of $20,000 installed. At present, the equipment is installed on a test site in an agrarian reform settlement near the town of Luiz Gonzaga, Maranhao, where it is producing approximately 10,000 kilograms of kernels weekly during the six-month babassu harvest season. The equipment and marketing operation are cooperatively administered, offering greater returns to producers than if they individually sell kernels they extract manually. Furthermore, once the small-scale oil presses are installed, this facility may achieve even greater financial viability. Charcoal and feed meal sales offer potential byproduct earnings that will further ensure the success of this enterprise.

In summary, the principal barriers to babassu palm development have included low native stand productivity, high bulk, and low unit value, competition from substitutes, rudimentary extraction techniques, and increasing restrictions in access to stands for fruit gatherers. Activities that attempt to address these problems in concert, through property rights definition in favor of extractivist households, community-based diversified processing facilities, and cooperative marketing may serve as models that can eventually reverse the decline that babassu industries have faced during the 1980s.

References

Amaral, J. Filho (1983). *A Extrato-Industria do Babassu no Maranhao.* Masters Thesis, Universidade Federal de Pernambuco.

Anderson, A., P. May, and M. Balick (1991). *The Subsidy from Nature: Palm Forests, Peasantry and Development on an Amazon Frontier.* New York: Columbia University Press.

Hecht, S., A. Anderson, and P. May (1988). The Subsidy from Nature: Shifting Cultivation, Successional Palm Forests, and Rural Development. *Human Organization* 47(l):25–35.

IBGE (1983). *Anuario Estatistics.* Brasil, Rio de Janeiro.

Mattar, H. (1979). *Industrialization of the Babassu Palm Nut: The Need for an Ecodevelopment Approach.* Conference on Ecodevelopment. Seminar Center of the German Foundation for International Development, Berlin.

May, P. (1986). *A Modern Tragedy of the Noncommons: Agro-Industrial Change and Equity in Brazil's Babassu Palm Zone.* Ithaca, NY: Latin American Studies Dissertation Series, Cornell University.

May, P. (1989). Local Product Markets for Babassu (Orbiqnya phalerata Mart.) and Agro-Industrial Change in Maranhao, Brazil. *Adv. in Econ. Bot.* (5)

May, P. (1990). *Palmeiras em Chamas: Transformaggo Agraria e Justir, a Social na Zona do Babassu.* Sao Luis, Maranhdo: EMAPA/FINEP/Ford Foundation.

MIC/STI (1982). *Mapeamento e Levantamento do Potencial das Ocorrgncias de Babaquais, Estados do maranhdo, Piaui, Mato Grosso e Golds.* Brasilia: Ministbrio da Indilstria e do dom-ercio.

Peters, C. (in press). Oligarchic Forest Potential for Development. In D. Nepstad and S. Schwartzman (eds.), *Extractive Resources in the Tropics* (proceedings of conference at National Wildlife Federation, Washington, D. C., November 1989). (To appear in Advances in Economic Botany.)

Pick, P. et al. (1985). *Babassu (Orbiqnya species): Gradual Disappearance versus Slow Metamorphosis to Integrated Agribusiness.* Report to the New York Botanical Garden Institute of Economic Botany, Bronx, New York.

Queiroz, L. (1989). Miséria e grandeza no reino do babassu. *Hoje* 9(51):8–10.

18

Nontimber Forest Products and Extractive Activities in the Middle Rio Negro Region, Brazil

JEAN-PAUL LESCURE, LAURE EMPERAIRE, FLORENCE PINTON, AND ODILE RENAULT-LESCURE
ORSTOM/INPA, Brazil

Extraction of nontimber forest products was an important source of income in the Brazilian Amazon up to the last world war (Santos, 1980; Homma, 1989). Currently, this industry is in a state of decline (Lescure and Castro, in press). Markets for things like rubber are depressed due to the commercialization of synthetic substitutes. This was the case of the latex of the Sapotaceae known as balata that was replaced by synthetics used to cover electric cables. Other products previously extracted from the wild like rubber (*Hevea spp.*) or guarana (*Paullinia cupana* var. *sorbilis*) are increasingly produced in plantation. On the regional level, the depopulation of rural areas and the concomitant growth of cities such as Manaus have contributed to the marginalization of extractive activities. Cultural factors such as the desire to relocate to areas where children can attend school has led to a global decline of extractivism. This can be seen clearly from the extractive activities/agriculture yield ratio that ranges from 58.7 percent in 1940 to 3.6 percent in 1990.

Paradoxically over the last five years, scientists and development planners have become increasingly interested in these types of activities. Research in the state of Acre, in collaboration with the rubber tappers' organization, has demonstrated the feasibility of these activities as a sustainable alternative to typical development activities (Allegretti and Schwartzman, 1987, Workshop of Belo Horizonte, May 1989: "Extrativismo na Amazônia: viabilidade econômica e dinámica populacional"). Model extractive reserves already exist in Acre.

Meanwhile, little is known about the different patterns of extractivism in Amazonia, particularly in the State of Amazonas. The aim of ORSTOM/INPA is to document the extractive activities in this state. This chapter focuses on extractivism in the middle Rio Negro Region and gives some preliminary results.

EXTRACTIVISM IN THE STATE OF AMAZONAS: AN INTRODUCTION

The latest data available from IBGE (Brazilian Institute for Geography and Statistics) show that the value of the extracted nontimber products was slightly over $7 million in 1987.

From an economic point of view, the income from extractivism represents less than 1 percent of state income, mostly due to the $9 billion from the industries of Manaus. Moreover, in terms of productivity, extractivism activities appear to be unproductive: the 570,000 rural inhabitants involved in extractivism produce less than 1 percent of the state income, as opposed to the 99 percent produced in the cities. If those economic data do not make an optimistic case for the future of extractivism, we have to keep in mind that extractivism remains immensely important from a sociological perspective, for it constitutes a big part of rural income.

Few products are in fact involved in those activities; from the total income that appears in the 1987 statistics, 69 percent was due to rubber (borracha, latex of *Hevea spp.*), 23 percent to Brazil nuts (castanha, seeds of *Bertholletia excelsa*), 4.5 to sorva (latex of *Couma spp.*), 2.1 percent to piassaba (fibers of *Leopoldinia piassaba*), 1.4 percent to copaíba oil (oleoresin of *Copaifera spp.*), and .1 percent to cipó titica (aerial roots of *Heteropsis spp.*). (See Figure 18-1.)

Each of these products shows a distinct trend between 1977 to 1989. Some, such as sorva, borracha, copaíba, and castanha are obviously decreasing. Piassaba seems to be increasing during the last five years.

THE MIDDLE RIO NEGRO REGION

The middle Rio Negro ranges from the mouth of the Rio Branco to the village of São Gabriel. The region is divided between the municipalities of Barcelos, Santa Isabel do Rio Negro and, for a part, São Gabriel da Cachoeira. Between those villages one finds tiny communities along the banks of the main river and its affluents.

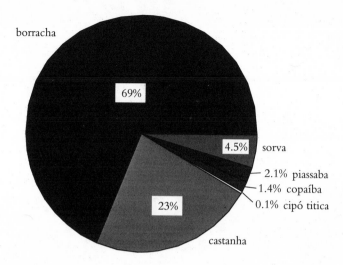

FIGURE 18-1 RELATIVE VALUE OF THE MAIN EXTRACTIVE PRODUCTS IN THE STATE OF AMAZONAS. (1987: TOTAL VALUE U.S.$7,153,492).

Access to the area is difficult and only the last village is linked to Manaus by a daily flight. It takes two days by boat to reach Barcelos and four days to reach Santa Isabel.

Climate is hot and wet. The rainfall ranges from 2,173 millimeters in Barcelos to 2,656 in Santa Isabel and 3,070 in São Gabriel within an east-west gradient. A drier season occurs from July to October but never leads to truly dry conditions. Mean temperature is 25° Celsius.

Included in the population data are Manaus and Novo Airão that are in the low Rio Negro. One can observe the dramatic increase of the Manaus population, currently estimated at 1.5 million. The populations of Barcelos and Santa Isabel are quite stable, while that of São Gabriel has been rapidly decreasing since 1980. Data demonstrate the very low density of the rural population, which is involved in extractive activities in the forest, ranging from .01 to .05 people per square kilometer.

FOREST PRODUCTS OF THE RIO NEGRO

Extractive activities focus mostly on sorva and piassaba. Both are exploited during the wet season, when the water level of the streams allows the transport of the products from the forest to the Rio Negro. During the dry season, people turn to rubber tapping, because the main local species, *Hevea spruceana,* grows on the flooded islands of the river.

The gathering of *Bertholletia excelsa* seeds occurs in May to June and is a casual activity.

Other products also occur in the region. *Astrocaryum jauari* is used in some parts of the Amazon to make hammocks. The oil of copaíba from the lower Rio Negro is consumed locally, as is the oil of andiroba, extracted from the seeds of *Carapa guianensis*.

Almost all of the piassaba comes from the Rio Negro as well as 40 percent of the sorva. For the other products, the importance of the region is little.

HOW THE MAIN PRODUCTS ARE COLLECTED AND PROCESSED

The Sorva

The method of collecting sorva depends on the species. *Couma macrocarpa* is always felled. A first ring is quickly made on the top of the log; informants say that it is done to impede the latex from flowing to the branches. Then rings are regularly made all along the trunk and the latex is collected below. Such a tree can give from 5 to 20 liters of sorva. This way of working is depleting the populations and the species is becoming rare.

The two species of *Couma, utilis* and *catingae,* as well as trees that belong to the genera *Parahancornia* and *Mucoa,* are tapped all along the trunk from the base to the top, while the worker climbs up. The latex is collected in a tin and transported to the camp where it is deposited in a hole (or *deposito*).

When enough latex is collected it is mixed with water in equal proportion in a 50-gallon tank. The mixture is heated until it coagulates. Another method mixes 80 liters of latex with 20 liters of water and 1 kilogram of salt; this way the latex coagulates without heat.

The coagulated latex is then cut into 10-by-10-centimeter blocks, and heated once more to make them hard. They are then deposited in a basket that will stay in the water before the embarkment of the product. One person can collect 20 to 50 kilograms of latex per day.

The Borracha

Tapping the *Hevea* trees (seringueiras) to extract the rubber (borracha) is a dry season activity, when the islands and the igapó are no longer flooded. Each collector alternatively works two footpaths called *estradas* that can have as many as 200 trees each. The trunk is incised slightly once or twice. Latex is then collected in a small tin, either bought in the city or made in the forest with spadice of *Euterpe catinga*.

Smoking the borracha to coagulate it is an abandoned technology. People now prefer to mix the latex with cassava juice (*tucupi*) in the proportion of 10 to 1. Coagulation takes place in just 1 to 2 hours. The coagulated latex is then cut into blocks and pressed to take the water out. The final product is called *sernamby virgem prensado* (SVP). One person can collect 10 to 20 liters of latex per day, which yields 4 to 9 kilograms of SVP.

The Piassaba

Piassaba collection is a rainy season activity, done when the water level is high enough to permit river access to the remote populations and easy transport of the product.

If the palm has never been worked, the first step consists of beating the trunk to make the dangerous animals (like poisonous snakes) fall off, and to free the fibers from one another. The collector then cuts the leaves to permit better access to the fibers, but leaves the 4 or 5 youngest ones so as not to kill the tree. A handful of fiber is then cut from the leaves that are younger than 5 years (the older are said to be brittle). The leaves are deposited on the ground between four stakes. When the pile is big enough to constitute a *fardo* (a cylindrical way to lash fibers), it is lashed with aerial roots of Araceae and transported to the camp. There, the *fardo* is opened and the impurities are taken out. The fibers are then lashed again. A person can collect 30 to 40 kilograms per day.

SOCIAL AND COMMERCIAL PATTERN OF EXTRACTIVISM

The usual pattern of relationships that exist between the different actors of extractivism is called *aviamento*. The basic scheme has three players: the extractor, the patrão, and the exporter. The extractor depends on a local dealer who also has the legal or historical extractive concession in the forest. This person, called the *patrão,* has a commercial relationship with exporters in Manaus.

The exporter lends a certain amount of money, in cash or in goods, to the *patrão*. This is then advanced to the extractor. This loan is known as the *rancho*. In turn the extractor, called the *freguês*, has to pay the *patrão* with his own production. Usually, the economic balance of the *freguês* is forever negative and the dependency of the *freguês* on the *patrão* cannot be broken. This makes the extractive activity unattractive for the forest people.

Prices also increase during the commercial process. Between the extractor and the export FOB price in Manaus, the value of sorva is mutiplied by 14.6, piassaba by 7.8, and castanha by 2.3. For the borracha the situation is different because the prices paid to the collector are controlled by the government.

The results of this pattern are that the extractors do not try to produce more. They prefer to become farmers to gain independence from the *patrão*. Through the cultivation of cassava and cassava flour production they will be able to eat all year and sell or barter the excess on the local market.

The second choice is to mix agricultural and extractive activities. However, during extraction time, the extractors will eat their own cassava flour, saving then a big part of the *rancho,* and they will spend a lot of time hunting and/or fishing.

CONCLUSION

In the middle Rio Negro, the main part of extractive activities appears to be sustainable from a biological point of view. The only destructive use is that of *Couma macrocarpa.* The dependency of the collector on the *patrão* seems to be the main factor contributing to the decline of extractive activities. The future of extractivism in this region will depend on socioeconomic solutions.

Acknowledgments

This study is part of the project "Extractivism in Central Amazonia, Viability and Improvement," worked out within the ORSTOM-CNPq cooperation agreement and supported by the UNESCO (MAB, grant SC–218–201–0), the French Ministry of Environment (SOFT, Grant 90049), and Conservation International.

References

Allegretti, M. H., and S. Schwartzman. 1987. *Extractive Reserves: A Sustainable Development Alternative for Amazonia.* Washington, D.C., World Wildlife Fund U.S., Project US–478, Mimeo.

Homma, A. K. O. 1989. A extracão de recursos natuarais renovaveis: o caso do extrativismo vegetal na Amazônia. Ph.D. Thesis, Vicosa University, UFB.

Lescure, J.P. and A. Castro. (in Press). L'extractivsme en Amazonie centrale. Apercu des aspects botaniques et economiques. In *Actes de l'Atelier sur l'amenagement et la conservation des Ecosystemes forestiers Tropicaux humides,* UNESCO, IUFRO, FAO, Cayenne, March 1990.

Santos, R. 1980. *Historia economica da Amazônia (1800–1920).* São Paulo, I. A. Queiros.

19

Colombian Palm Products

Rodrigo G. Bernal

Instituto de Ciencias Naturales, Universidad Nacional de Colombia

Most of the 247 species of native palms in Colombia are used in some way at the local level. At least 20 species provide products that are marketed to some degree, and three of these form the basis of significant industries: *Euterpe oleracea* (palm hearts), *Leopoldinia piassaba* (piassava fiber or chiquichiqui), and *Phytelephas macrocarpa spp. schottii* (tagua or vegetable ivory). This chapter focuses on establishing appropriate programs to manage these three species and highlights the urgent need to develop projects to exploit other species as well.

Palms are one of the three most useful families of plants known and one of the 15 largest families in Colombia, where they are by far the most widely used among the rural population. From the flexible stems of the matamba (*Desmoncus cirrhiferus*), used to construct cradles, to the corpulent stipes of the barrigona (*Dictyocaryum lamarckianum*), which are used as coffins, palms are forever present in human life in the jungle. They offer a wide range of products such as foods, beverages, clothing, construction materials, tools and household wares, weapons, and jewelry.

Studies carried out in the northwestern part of the Colombian Amazon show that palms dominate the vegetation, and that they are the most important forest resource for the local communities (Palacios, 1989). Galeano (1991) has noted that palms may be a valuable resource for the Colombian economy, ensuring the conservation of the forests. Thus, palms are perhaps the first family that must be considered in an evaluation of nontimber commercial forest products. In fact, in their evaluation of one hectare of forest near Iquitos, Peters et al. (1989) found 4 species of palm among 11 species of trees producing marketable fruit or latex. Of the trees with a diameter larger than 10 centimeters, 60 percent were from these four palm species, and their fruit comprised 56 percent of the total value of the nontimber products in that hectare.

Over half of the 247 native palm species currently recorded in Colombia have known uses. Products from at least 20 wild palm species are marketed either at the local level or on a broader scale. These species, along with their respective products, are summarized in Table 19-1. It is quite likely that other species not included on this list are also marketed on a very small scale at the local level. Because of the local nature of the commerce in almost all these products it is difficult to estimate their total production value.

This chapter does not cover cultivated species of palm, such as the coconut (*Cocos nucifera*), the chontaduro or peach palm (*Bactris gasipaes*), or the African oil palm (*Elaeis guineensis*).

Fruits from at least 5 species are sold in different regional markets in Colombia. Those from the corozo or mararay (*Aiphanes caryotifolia*) are common in several markets in the coffee region and can even be found in supermarkets in Medellín. Sugar candies with a corozo pit embedded in the center are common in several cities throughout the Andean region. The fruits from the tucumá (*Astrocaryum vulgare*) and canangucha or aguaje (*Mauritia flexuosa*) are sold fresh in food markets in Leticia, on the Amazon River. Fruits from the lata (*Bactris guineensis*) are common in the market in Cartagena and are used to make corozo juice, which is sold in many restaurants in that city. Fruits from the naidí (*Euterpe oleracea*) are sold in the market of Buenaventura, and are used to make a juice.

In the department of Chocó, baskets made with fibers from the antá (*Ammandra decasperma*) and the güérregue (*Astrocaryum standleyanum*) are both sold locally, with the latter fetching high prices in arts and crafts stores in Bogota. Likewise, hammocks and hats made of cumare (*Astrocaryum chambira*) are sold in town stores in the Amazon and the Orinoco regions, as well as in the stores of large cities. The use of *Astrocaryum* fibers from the young palm leaves can be destructive, often causing the death of the plant.

The leaves of the palma escoba [broom palm] (*Cryosophila kabreyeri*) are sold in the local markets near the Sinú River, where they are used to make brooms. The leaves of the palma amarga [bitter palm] (*Sabal mauritiiformis*) are a valuable roofing material that is very popular in the Caribbean coastal region, where the palms are zealously cared for and regularly harvested.

Moreover, two species are used in Bogotá as foliage in flower arrangements for funerals: *Geonoma lindeniana* from the eastern range of the Colombian Andes and *Phoenix roebelenii*, originally from the peninsula

TABLE 19-1 *Principal Marketed Products from Wild Colombian Palms*

Common Name	Species	Organ	Products	Region
Mararay	*Aiphanes caryotifolia*	Fruits	Fruits	Andean
		Seeds	Candy	
Antá	*Ammandra decasperma*	Petiole	Baskets	Pacific
Cumare-Tucumá	*Astrocaryum vulgare*	Leaves	Hammocks, hats	Amazon
		Fruits	Fruits	
Güérregue	*Astrocaryum standleyanum*	Leaves	Baskets	Pacific
Lata	*Bactris guineensis*	Fruits	Juice, Fruits	Caribbean Coast
Palmas de ramo	*Ceroxylon (3 spp.)*	Leaves	Palm Sunday fronds	Andean
Caña de San Pablo	*Chameadorea pinnatifrons*	Plants	Ornamental	Andean
Palma escoba	*Crysophila kalbreyeri*	Leaves	Brooms	Caribbean Coast
Chicón	*Dictyocaryum lamarckianum*	Seeds	Handicrafts	Andean
Naidí	*Euterpe oleracea*	Meristems	Palm hearts,	Pacific
		Fruits	fruits	
Palmicho	*Euterpe precatoria*	Inflorescence	Floral	Andean
	Geonoma lindeniana	Leaves	industry	
			Foliage	Andean
Chiqui-Chiqui	*Leopoldinia piassaba*	Fibers	Brooms	Orinoquía
Aguaje-Moriche	*Mauritia flexuosa*	Fruits	Fruits	Amazon
Jicra	*Manicaria saccifera*	Bracts	Hats/bags	Pacific
Tagua	*Phytelephas macrocarpa spp. schottii*	Seeds	Handicrafts	Andean
Palma amarga	*Sabal mauritiiformis*	Leaves	Thatch	Caribbean Coast
Palma de vino	*Scheelea butyracea*	Leaves	Palm Sunday branches,	Andean
		Sap	Wine	

of Indochina, but now cultivated in Colombia. Every day, hundreds of leaves are used for this purpose in Bogotá. Research has not been carried out to study the impact the use of their leaves has on the *Geonoma lindeniana* populations.

The still unopened leaves of the *Ceroxylon* species are sold by the thousands every year in many cities and towns of Colombia for Palm Sunday celebrations. This use is destructive and causes the death or retardation of the species in this genus (Bernal, 1989b), most of which are endangered (Bernal, 1989a). In Bogotá, and perhaps other cities as well, thousands of palma de vino [wine palm] (*Scheelea butyracea*) leaves are sold for the same purpose. This species rapidly regenerates in areas that have been disturbed; thus, their use for this purpose does not seem to pose an imminent danger. Nevertheless, the *Scheelea* leaves are less esteemed than those of the *Ceroxylon* and are sold at a lower price.

The inflorescences of palmicho (*Euterpe precatoria*) are used by florists and sold in some cities. The fibrous peduncular bracts of the jicra (*Manicaria saccifera*) are used in Chocó to make hats and bags and are common in arts and crafts stores.

The seeds of the chicón (*Dictyocaryum lamarckianum*) and of the tagua (*Phytelephas macrocarpa ssp. schottii*) are used in the city of Chiquinquirá to create many handicraft objects, such as small animal figurines, rosaries, and the like.

The sap of the palma de vino [wine palm] (*Scheelea butyracea*) is extracted by cutting the palm, and then fermenting it until it becomes a whitish wine. It is sold in the dry zones where this palm abounds. Although this use is destructive, it does not appear to threaten the populations of the species, which grow quite well in open areas.

Finally, the *Chamaedorea pinnatifrons* plants are removed from the forest by campesinos and sold in street-side markets in some cities.

Three products from uncultivated Colombian palms are marketed more broadly and should be examined in more detail. They are palm hearts, vegetable ivory, and piassava fiber.

PALM HEARTS

Palm hearts are the young leaves, at different stages of development, of several palm species. Most Colombian palm heart production comes from the *Euterpe oleracea*. This palm, which is called naidí in the Pacific coastal region, has been known in Colombia as *Euterpe cuatrecasana* (Dugand, 1951; Patiño, 1977) and in Ecuador as *Euterpe chaunostachys* (Pedersen and Balslev, 1990). Recent studies show, however, that it is the same species as the *Euterpe oleracea* that grows in Brazil (G. Galeano, pers. comm.).

This palm is very common throughout the lowland flood areas along the rivers of northern South America. In Colombia, it is found west of the Andes as well as in the Middle Magdalena River Valley. The greatest stands of this species are in the southern part of the Pacific region, where they form extensive communities, known locally as naidizales or naidi groves, from where the marketed palm hearts are currently harvested.

Euterpe oleracea is a caespitose palm that forms as many as 10 trunks, growing to 15 meters in height and 10 or 15 centimeters in diameter. It has pinnate leaves whose sheaths form prominent crownshafts of reddish color.

The palm heart is obtained by cutting the palm and removing the crownshaft, in which the heart is found; later, the sheaths of the crownshaft are removed in order to extract the palm heart found inside. The palm hearts are canned directly in the areas of production.

The exploitation of the palm is partially destructive because the palm must be felled to extract the palm heart. Nevertheless, because it is cespitose, the plant does not die when its stem is cut; instead, other stems can develop. This characteristic is highly advantageous for species management. In single-stem species, such as *Euterpe edulis,* palm heart exploitation without adequate management has decimated the populations of this species at an alarming rate (Nodari and Guerra, 1986). Even though the cespitose nature of *Euterpe oleracea* offers the possibility of sustainable exploitation, it does not ensure it per se, and the method of exploitation used is not always ideal for guaranteeing the plant's survival.

As of May 1991, six companies operate *Euterpe* palm heart canning factories in the Pacific coast region of Colombia, from the mouth of the San Juan River, just north of Buenaventura, to the Ecuadoran border. One is at the mouth of the San Juan River itself, another in the village of Guapi in the department of Cauca. Two are in the village of Salahonda in the department of Nariño, and two are in the port city of Tumaco in the same department.

Although palm heart exploitation in the Pacific coast began in 1974 (Patiño, 1977), the statistics on recent exports of this product from Colombia date back only as late as 1984. As of that year, palm heart exports have constantly grown (see Figure 19-1) with the possible exception of 1985, for which there are no statistiscs (DANE, 1986a, 1986b, 1988, 1990; INCOMEX, 1990). Although the statistics do not specify the origin of the exported palm hearts, the majority are from the *Euterpe oleracea;* only one company has chontaduro (*Bactris gasipaes*) palm heart

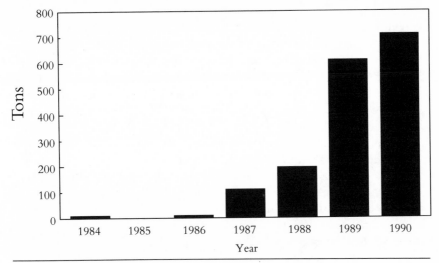

FIGURE 19-1 PALM HEART EXPORT FROM COLOMBIA 1984–1990.

SOURCE: INCOMEX

plantations, and these did not go into production until recently (Velasco, 1989). Current *Bactris* palm heart exports account for only 10 percent of total palm heart exports.

Of the countries importing palm hearts from Colombia, France is by far the most significant. In the period from January 1989 to August 1990, France accounted for almost four-fifths of total exports of palm hearts, followed by the United States, Spain, Holland, Belgium, Luxemburg, and Japan.

There have not been any studies conducted in Colombia on the long-term effects of palm heart exploitation, and differences of opinion exist with regard to the impact of the industry on natural populations. The exploitation first began before a study of the dynamics of the naidí palm groves could be undertaken. Thus, in 1977, the Colombian Institute of Natural Resources and the Environment (INDERENA) prohibited palm heart exploitation. Later, in 1978, INDERENA permitted it on a regulated basis. After carrying out a short-term study (Tibaquira, 1980), INDERENA concluded that palm heart exploitation helped the development of the naidí palm groves. However, this is contradicted by local informants in the southern Colombian department of Nariño and the northern Ecuadorean province of Esmeraldas, who claim that these palm populations have been severely reduced by the exploitation. Patiño (1977) said that *Euterpe oleracea* was being wiped out in the Guapi region as a result of the commercial exploitation that began in 1974. At present the companies that market palm hearts are carrying out studies to determine the impact of exploitation on the palm groves. Because the studies are being carried out by the companies themselves, they must be carefully evaluated.

Under the current method of exploitation, the palm hearts are purchased from native collectors, without any effective control over their management of the palm colonies. Each collector attempts to obtain the greatest number of palm hearts, regardless of the damage that over-collection may inflict on the resource. This may be the reason that the potentially sustainable exploitation becomes destructive. CORPONARIÑO (1989) has recognized the need to train the workers of these companies in the methods and systems of handling and using this resource. A system of extraction through small community businesses could encourage more rational care of the palm and prevent the exhaustion of the resource. This could also be an option to exploit the palms in areas where the naidí stands are too few to justify large-scale companies.

TAGUA

Tagua or vegetable ivory is a very hard cream-color endosperm that looks similar to ivory. It comes from several species of palms that grow wild in Panama, Colombia, Ecuador, Peru, and northwest Brazil. The most important commercial genus is *Phytelephas*.

Taguas vary in size from small acaulescent palms with a subterranean stem and a rosette of large pinnate leaves to medium-sized palms up to 10 or 12 meters in height. They are doecious and produce large fruits that are grouped into spherical heads. The fruits are obconical, growing up to 15 centimeters in diameter, with an epicarp adorned with spiny ligneous projections. Each fruit contains from 4 to 10 irregularly-shaped seeds that grow up to 7 centimeters. Each seed is covered by a hard ligneous endocarp that turns fragile when dry. At first, the endosperm is watery, but it turns gelatinous as it matures, and then hardens to form an ivory-like material that constitutes the tagua or vegetable ivory.

Tagua palms grow in humid lowland forests, especially in alluvial plains, where they frequently form huge palm groves called taguales. The mature seeds fall to the ground, where they are dispersed mainly by rodents such as the agouti (or ñeque) (*Dasyprocta punctata*) (Barford, 1991) and by gravity or river flooding. These are the seeds that are collected and sold. With the exception of some areas near the Santiago

River in the Pacific coastal region of Ecuador (Acosta Solis, 1944), tagua palms have never been cultivated, and all the tagua production comes from wild populations.

Shortly after it was discovered, tagua was recognized as a useful raw material in the production of several objects, particularly buttons, and within a few years it became an important forest resource that played a significant role in the economies of Ecuador and Colombia until the 1940s, when it was replaced by plastics (Acosta Solis, 1944; Barford, 1989). Thereafter, the tagua button industry barely survived in Ecuador (Barford et al., 1990) along with the economically insignificant tagua handicraft industry.

Tagua first appears in Colombian statistics for 1840 and 1841, in which it constitutes a small fraction of exports for those years. Nevertheless, its trade continued to grow without interruption, and between the 1860s and 1891 it was making a modest but constant contribution to Colombian exports (Ocampo, 1984). Tagua reached its peak in the nineteenth century in the period between 1875 and 1878, when it accounted for 3.1 percent of Colombian exports (see Figure 19-2) (Tovar Zambrano, 1989). Then, it fluctuated slightly, peaking in 1911 at 3.4 percent. Thereafter, tagua production declined until the product disappeared al-

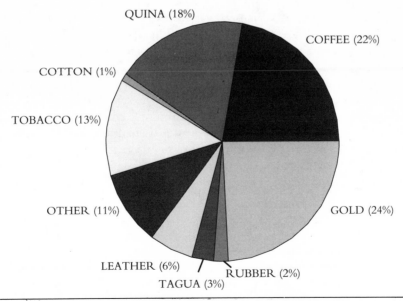

FIGURE 19-2 TAGUA AMONG COLOMBIAN EXPORTS 1875–1878.

SOURCE: Tovar Zambrano. 1989.

together from international markets in the 1950s (Domínguez and Gómez, 1989). Tagua from remote areas of the Pacific Coast was gathered and carried in small boats directly to Panama. Thus, it may never have been recorded in statistics.

Despite the loss of the international tagua market, the tagua handicraft industry never disappeared. Around 1931, a tagua handicraft factory was set up in the Andean city of Chiquinquirá, which received tagua from the Middle Magdalena River Valley (*Phytelphas macrocarpa ssp. schottii*). This factory initiated the tagua handicraft industry, which still survives in that city.

The Chiquinquirá tagua industry consists of family-owned and run businesses. Until 1987, the tagua handicraft industry flourished, with approximately 100 families depending on it for their livelihoods. Since that time, the industry has declined and only 14 families now depend on it. Each factory consumes between 250 and 1,000 kilograms of tagua each month. If the minimum monthly figure of 250 kilograms is used, the total monthly consumption obtained is 3.5 metric tons. This means that a few years ago when 100 families worked with tagua, consumption may well have been over 25 metric tons per month. However, this figure is only an approximate estimate, based on local information.

Tagua handicrafts are produced on lathes; there is no tradition of carving tagua in Chiquinquirá. Approximately 1,000 different designs are produced, most of which are variations on just a few themes. The most common include chess pieces, small bowls, drinking glasses, and animal figurines. The sale of these handicrafts is totally restricted to Chiquinquirá.

The recent increase in international demand for tagua, used in button production, once again opens the doors for the sustainable exploitation of this product in Colombia. Local community management of tagua production, which is already under way in Ecuador, offers these communities the opportunity to practice forest conservation. At present, the first steps toward tagua production are being taken in Colombia.

PIASSAVA OR CHIQUICHIQUI

Piassava or chiquichiqui is a strong, dark brown fiber, between 1 and 1.5 meters long, produced by the leaf sheaths of the chiquichiqui palm (*Leolpoldinia piassaba*) that grows in the basins of the Rio Negro and the Orinoco River in Colombia, Brazil, and Venezuela.

Chiquichiqui, known in Brazil as piassava, is a solitary palm that reaches a height of approximately 10 meters, and has a crown of numerous pinnate leaves that grow to 4 meters in length. The leaf sheath is composed of numerous fibrovascular bundles, joined at the base in bands 3 centimeters wide. The latter decay over time, turning into individual fibers as much as 1.5 meters in length. The leaf sheaths cover the palm and the fibers hang freely.

The chiquichiqui palm grows wild in poor and sandy soils, mainly associated with black water rivers, where it forms almost exclusively single-species palm groves of several hectares in area. It is from these palm groves that the fiber is collected and marketed as piassava. The fiber is mainly used to manufacture ropes and brooms. The ropes, the diameter of which may exceed 15 centimeters, are light, resistant, and particularly appropriate for marine use because of their resistance and bouyancy (Putz, 1979) as well as the fact that they decompose when left to drift in the ocean, a characteristic that would reduce the navigation problems that synthetic lines may cause.

Chiquichiqui brooms are as resistant as those made from synthetic materials. However, in recent years they have lost popularity even though their product quality is not inferior. As with the ropes, chiquichiqui brooms have the advantage over their synthetic counterparts because they are biodegradable.

There are no statistics on chiquichiqui or piassava fiber exports from Colombia in recent years, although there may be small-scale exports that are not recorded in the yearbooks. Current fiber production in Guainía department is between 820 and 850 tons per year (Alba Páez, 1990). As of May 1991, the price of the fiber in Bogotá was at 270,000 pesos or $450 (U.S.) per ton. An undetermined portion of the fiber production goes to broom factories in several Colombian cities. A factory in Medellín uses 1.5 tons per month to produce between 600 and 1,200 brooms, which are sold at a unit price ranging from 500 to 565 Colombian pesos, the equivalent of $0.83 to $0.93.

The geographic and ecological distribution of chiquichiqui palm, along with the distribution of human populations of principally indigenous communities in Guainía, make this plant ideal for establishing a community management system. In fact, such a project was begun in Guainía in 1990 (Alba, 1990) and is now in full swing.

OTHER SPECIES

In addition to the species currently being marketed, there are other palms from rain forests, humid forests, and other ecosystems that could provide the basis of extractive systems. Particularly noteworthy among these are the milpesos (*Oenocarpus bataua*) and other *Oenocarpus* species, the táparo (*Orbignya cuatrecasana*), the palma de vino [wine palm] (*Scheelea excelsa*) and the corozo palm, also called the tamaco (*Acrocomia aculeata*).

The milpesos palm, also known in Colombia under other names, such as seje, is referred to in the botanical literature as *Jessenia bataua*. It is an oleaginous species that produces an oil similar in terms of its chemical and organoleptic properties to that of the olive (Balick, 1981; Balick and Gerschoff, 1981). This palm has a single trunk of up to 20 meters in height, with enormous pinnate leaves and huge panicles of ellipsoid fruits, growing up to 4.5 centimeters in length.

This species is broadly distributed in northern South America. It is found in all the humid lowland forests of Colombia: in the Pacific region, the Middle Magdalena River Valley, and the Amazon region, as well as the gallery forests of the Orinoco region. It is extremely abundant in the Colombian Pacific region, and in some areas is the most common tree species (Faber-Langendoen and Gentry, 1991). This species has been noted by Balick as promising for cultivation (1988), and because of its great abundance in forests in some areas, it could have great potential as the basis of an extractive industry.

A program for milpesos oil production through community businesses would include the installation of small presses in some villages in which the program is developed, as well as the supply of palm-climbing equipment, along with an active product marketing process. The great abundance of this species near areas of human inhabitation in the department of Chocó makes this region a good pilot area to begin an extractive production program with this species. Although other species of *Oenocarpus* produce fewer fruits, they still are significantly oleaginous and could contribute to *O. bataua* production. In fact *O. mapora* is one of the four palms mentioned for its market value by Peters et al. (1989) in his study of a hectare in Iquitos.

Táparo is an acaulescent palm with enormous leaves that sprout from the ground level and huge fruits of up to 15 centimeters in diameter, which are among the largest fruits of American palms. This species is endemic in the Colombian Pacific coastal region, where it frequently forms large palm groves called taparales in under-forest flood areas. The seeds of this palm consist of as much as 48 percent oil, thus it has been considered by Patiño (1977) as the most important oleaginous palm of the Colombian Pacific area. However, this same author noted that the seed displays the impediment of a thick endocarp that complicates the extraction of its nuts. According to Cuadros (Patiño, Appendix IV, 1977) there is a market for this palm oil in the San Juan River region.

Nevertheless, this species is not included in Table 19-1 because its trade is of a local nature and because no one knows if this trade continues. The prospect of producing *Orbiqnya cuatrecasana* on a larger scale needs to be studied in detail.

The palma de vino [wine palm] (*Scheelea excelsa*), sometimes referred to as *Scheelea magdalenica,* with which it is apparently cospecific (Galeano and Bernal, 1987) is a solitary corpulent palm that produces large panicles of yellow fruit whose seeds are rich in oil. This species is quite abundant in the lower Magdalena River Valley, where it forms extensive high-density palm groves. The seeds consist of between 50 percent and 60 percent oil, of a kind similar to that of the coconut (*Cocos nucifera*) (Dugand, 1958). This oil was the basis for an industry established in the city of Barranquilla, at the mouth of the Magdalena River on the Colombian Caribbean coast, where the seeds purchased throughout the entire lower Magdalena region were processed. Although the industry failed, apparently because of problems in the seed supply system, there are still extensive populations of this palm in pastures, and its production in combination with livestock grazing should be considered.

The corozo or tamaco (*Acrocomia aculeata*) is common in many tropical dry forest areas in Colombia, where at times it is abundant, regenerating easily in disrupted areas. Oil extraction from its fruits, both from the mesocarp as well as the seed, was once common, although now it is no longer a practice, in the department of Antioquia (Galeano and Bernal, 1987). Lleras (1985) has indicated that the genus *Acrocomia* may be the most important of all oleaginous palms and considers it ideal for cultivation, an endeavor for which a germ plasm bank has already been created (Coradin and Lleras, 1986). The large populations of this palm found in certain dry inter-Andean areas, the Caribbean coastal plains, and especially in the north of the extensive Llanos Orientales (or eastern plains), could become the basis for small-scale industries.

There are other species that, despite having little or no current use, could become the basis for small-scale extractive industries if assessments of their possible use were conducted. Three species from western Colombia are good examples: the güérregue (*Astrocaryum standleyanum*), the pángana (*Raphia taedigera*), and the jingapá (*Bactris sp.*).

The güérregue (*Astrocaryum standleyanum*) is a spiny palm widely dispersed west of the Andes in Colombia. Because of its oleaginous seeds, this palm has been included by Patiño (1977) among the species of immediate economic-industrial potential.

The pángana (*Raphia taedigera*) is a cespitose palm of monocarpic stems and gigantic leaves that grow to 15 meters in length. It forms extensive homogenous communities in permanently flooded places along the lower Atrato River in the departments of Chocó and Antioquia (Galeano and Bernal, 1987). Although the palm is used very little in this area, other *Raphia* species are greatly utilized by the local populations of the west coasts of Africa (Profizi, 1986) and the raffia fibers are well known on the international market. Thus, the prospects of this species need to be assessed.

The jingapá palm (*Bactris* sp.) is a *Bactris* species that does not yet appear to have been described (Galeano and Bernal, 1987). It is widely dispersed throughout the lowland flood areas west of the Andes, where at times it is quite abundant. This is a cespitose palm that grows from 10 to 12 meters in height and from 10 to 15 centimeters in diameter. Its wood, as with that of other species of its genus, is very hard and could lend itself to many uses. Because of its cespitose character and its abundance in flooded areas inappropriate for agricultural use, this palm could become the basis for small extractive industries should the studies of its wood confirm the potential that preliminary observations suggest.

Finally, the great ornamental value of palms should be highlighted. The growing interest in introducing palm species into cultivation in Colombian cities and other countries offers additional opportunities for this group. The propagation and marketing of native palm species could become an additional source of revenue. At the same time, pressure would be eased on the wild populations of these species (for example, *Chaemaedorea pinnatifrons*). Likewise, such an endeavor would assure the survival of the endangered species.

Acknowledgments

I wish to thank Conservation International for its support. Also, thanks to Gloria Galeano, Edgar Linares, and Diego Restrepo for their valuable information. Thanks to Angélica Peñuela for the figures that illustrate this chapter.

References

Acosta Solis, M. (ed.) 1944. *La tagua*. Ecuador: Quito.

Alba Páez, D. E. 1990. *Segundo informe de avance de comercialización de fibra de chiquichiqui Guainia*. Typed report. Corporación Colombiana para la Amazonia, Bogotá.

Balick, M. J. 1981. Jessenia bataua and Oenocarpus species: Native Amazonian Palms as New Sources of Edible Oil. In E. H. Pryd, L. H. Princen, and K. D. Mukherjee (eds.), *New Sources of Fats and Oils*. Champaign, IL: Oil Chemists Society.

Balick, M. J. 1988. *Jessenia and Oenocarpus: Neotropical Oil Palms Worthy of Domestication.* FAO Plant Production and Protection Paper #88. FAO. Rome.

Balick, M. J., and V. S. N. Gerschoff. 1981. Nutritional Evaluation of the Jessenia Bataua Palm: A Source of High Quality Protein and Oil from Tropical America. *Econ. Bot.* 35(3):261–71.

Barford, A. 1989. The Rise and Fall of Vegetable Ivory. *Principes* 33(4):181–90.

Barford, A. 1991. A Monographic Study of the Subfamily Phytelephantoideae (Arecaceae). *Opera Botanica* 105:5–73.

Barford, M. J., B. Bergmann, and H. B. Pedersen. 1990. The Vegetable Ivory Industry: Surviving and Doing Well in Ecuador. *Econ. Bot.* 44(3):293–300.

Bernal, R. G. 1989a. Endangerment of Colombian Palms. *Principes* 33(3): 113–28.

Bernal R. G. 1989b. Las palmas de cera del Quindío. *Lámpara* (Bogota) 27(110): 23–29.

Coradin, L., and E. Lleras. 1986. Coleta de germoplasma de macaúba—situaçao atual. *Useful Palms of Tropical America* 2:5–6.

CORPONARIÑO (Corporación Autónoma Regional para el Desarrollo de Nariño). 1989. *El naidizal del Departamentol de Nariño.* Typed report. CORPONARINO, Pasto.

DANE (Departamento Administrativo Nacional de Estadística). 1986a. *Anuario de Comercio Exterior 1984.* DANE, Bogotá.

DANE. 1986b. *Anuario de Comercio Exterior 1985.* DANE. Bogotá.

DANE. 1988. *Anuario de Comercio Exterior 1986.* DANE. Bogotá.

DANE. 1990. *Anuario de Comercio Exterior 1988.* DANE. Bogotá.

Domínguez, C., and A. Gómez. 1989. *La economía extractiva en la Amazonia colombiana 1850–1930.* Tropenbos-Corporación Colombiana para la Amazonia, Bogotá.

Dugand, A. 1951. Palmas nuevas o notables de Colombia. *Rev. Acad. Colombia* 8:385–96.

Dugand, A. 1958. Una palma nueva *Scheelea* del Bajo Magdalena. *Mutisia* 1(26):1–6.

Faber-Langendoen, D., and A. Gentry. 1991. The Structure and Diversity of Rain Forests at Bajo Calima, Chocó Region, Western Colombia. *Biotropica* 23(1):2–11.

Galeano, G. 1991. *Palmas de la region de Araracuara: Estudios en la Amazonia Colombiana.* Tropenbos, Bogotá.

Galeano, G., and R. Bernal. 1987. *Las palmas del Departamento de Antioquia: Región Occidental.* Universidad Nacional de Colombia, Bogotá.

INCOMEX (Instituto Colombiano de Comercio Exterior). 1990. *Sistema estadístico de comercio exterior SECE.* Computer printout. INCOMEX, Bogotá.

Lleras, E. 1985. Acrocomia, um gênero com grande potencial. *Useful Palms of Tropical America* 1:3–6.

Nodari, R. O., and M. P. Guerra. 1980. O palmiteiro no sul do Brazil: situaçao e perspectivas. *Useful Palms of Tropical America* 2:9–10.

Ocampo, J. A. 1984. *Colombia y la economía mundial 1830–1910*. Ed. Siglo XXI, Bogotá.

Palacios, P. A. 1989. *Aspectos de la utilización del bosque maduro en la Amazonia colombiana*. II Simposio Colombiano de Etnobotanica. Popayán.

Patiño, V. M. 1977. Palmas oleaginosas de la costa colombiana del Pacífico. *Cespedesia* 6(23–24):131–260.

Pedersen, H. B., and H. Balslev. 1990. *Ecuadorean Palms for Agroforestry*. AAU Reports 13.

Peters, C. M., A. H. Gentry, and R. O. Mendelsohn. 1989. Valuation of an Amazonian Rainforest. *Nature* 339:655–56.

Profizi. J. P. 1986. Biologie et modes de gestion des marecages à *Raphia hookeri* Mann et Wendland au sud-est du Benin. *J. Agric. Bot. Appl.* 33:49–58.

Putz, F. 1979. Biology and Human Use of *Leopoldinia piassaba*. *Principes* 23(4):149–56.

Tibaquira, C. L. 1980. *Potencial de los bosques de palma naidi en la costa sur del Pacifico colombiano (Cauca, Nariño)*. Mimeographed report. INDERENA. Bogotá.

Tovar Zambrano, B. 1989. *La Economia colombiana (1886–1922)*, Vol. 5. In Tirado Mejia, A. (ed.), *Nueva Historia de Colombia*. Bogotá: Planeta.

Velasco, F. A. 1989. En Colombia, el palmito de chontaduro (Bactris gasipaes). *Serie Técnica Pejibaye* 1(2):13–14.

20

The Economic Botany of Ecuadorean Palms

HENRIK B. PEDERSEN AND HENRIK BALSLEV
Botanical Institute, University of Aarhus, Denmark

Lying on the equator, on South America's Pacific coast, Ecuador is naturally divided into three biogeographic zones: the eastern lowlands (which are part of the upper Amazon Region), the coastal plains, and the mountains. Ecuador covers some 275,000 square kilometers, making it one of the continent's smallest nations.

Yet, despite its size, Ecuador harbors an estimated 20,000 species of vascular plants (Balslev, 1988), which are distributed in 26 different life zones (Cañadas Cruz, 1983). Palms are a conspicuous component of the flora occurring in these zones, except in dry areas and zones above 3,000 meters. To date, a total of 129 native palm species in 34 genera as well as 15 introduced species in 13 genera have been recorded (Balslev and Barford, 1987; Balslev, 1990).

For the indigenous people of lowland Ecuador, palms constitute one of the most significant families of plants. Palms supply food, beverages, medicine, fibers, and thatch, as well as materials for construction and hunting implements.

The commercial value of palms also is increasingly important to indigenous people. For example, Descola (1989) concluded that Achuar Indians select sites for establishing new villages based on the proximity of these sites to dense stands of two palm species that the Achuar use in trade: kinchuk (*Aphandra natalia*), which supplies fibers; kunkuk (*Jessenia bataua*), favored for its oil fruits; and the lauraceous ishpink (*Ocotea quixos*), which is valued for its cinnamon-flavored cupola. In Ecuador, the value of products extracted from natural palms extends into the marketplace. Palm wood is used to fashion furniture and keepsakes. Edible palm hearts and various fruits are sold in city markets; some are exported. Palm fibers are used to make brooms and hammocks. Palm seeds and the mesocarp (or edible flesh) of fruit provide oils. And vegetable ivory (mature, hardened endosperm—or seeds—of the fruit) is used to make such items as buttons.

The purpose of this chapter is to summarize the commercial use and management of Ecuadorean palms. Its content is based on information gathered from literature on the subject and on personal observations. (When vernacular names from the literature are cited, we have provided up-to-date nomenclature or possible interpretations in brackets.)

EARLY COMMERCIAL EXPLOITATION OF PALMS

The commercial exploitation of palms in Ecuador has a long though not particularly strong tradition. In his *Historia del Reino de Quito* (1789), Juan de Velasco mentions several products that apparently had commercial value during his day. Wax, obtained from the wax palms [*Ceroxylon* spp.], was called "cera de valles" and is described as "very useful for making good business." The wax probably was used to make candles, a practice that ended only recently. The fruits or seeds of an iriarteoid palm, called pona [probably *Iriartea deltoidea*], were sent all the way to Lima, where it was "a custom to have it embedded in gold, just because of the beauty of it." Meanwhile, fruto del "corozo" [vegetable ivory from *Phytelephas aequatorialis* and other species] was used to make "figurines, carvings of saints, and other curiosities." Such carvings, dating back to the 1700s, can still be bought in Quito's antique shops, selling for $200 to $500 each. Juan de Velasco also mentions other palm products, several of which have the same vernacular names and/or uses today.

In Amazonian Ecuador, some palm products have been used—and, to a certain degree, still are used—in trade between different tribes or different subgroups within tribes. Karsten (1935) and Descola (1989) both refer to blowguns made from chonta [*Bactris gasipaes*] and from tuntuam [*Iriartea deltoidea*], among other palms, as products used in intertribal trade. According to Descola, Achuar Indians are well known blowgun makers and supply their creations to both the Quichua Indians of Canelos and the Shuar Indians. Meanwhile, Karsten described that, in trading with the Achuar, the Canelos-Quichua offered blowgun darts, made principally from the leaf rachis of inayu [*Maximiliana maripa*]. As the darts were easy to make, the main reason for the trade was the near absence of inayu in Achuar territory (Karsten, 1935).

In the past, vegetable ivory was undoubtedly one of the most commercially important palm products. Ecuadoran statistics show that its export began in 1870. That first shipment weighed 1,100 tons, most of it destined to become buttons. By 1909, the export had reached 18,000 tons per year and ranked as the country's second most important exported product, after cacao. "Panama hats," made from the leaves of *Carludovica palmata* (Compania, *Guia del Ecuador,* 1909), were the third most exported product. The overseas sale of vegetable ivory decreased dramatically at the outbreak of World War II and became almost nonexistent after the war, as synthetic materials came into use for making buttons (Acosta Solis, 1944; Barford, 1989, 1991b). According to Barford et al. (1990), the demand for vegetable ivory has recently risen.

AN OVERVIEW OF ECONOMICALLY SIGNIFICANT ECUADORAN PALMS

Although a few native palm species are cultivated to a limited extent in Ecuador, it is primarily palms growing in the wild that are exploited. To date, we have observed that commercial products are derived from 12 wild-growing species in 10 genera. A number of these products, however, must be classified as relatively insignificant economically on both national and local levels.

The following information describes various native palm genera in Ecuador that are economically valuable. Also, it includes two species known only in cultivation: *Bactris gasipaes* and *Parajubea cocoides.* Though both are from the Americas, their exact origin is unknown. It does not include the most economically significant palm in Ecuador, the introduced African oil palm (*Elaeis guineensis*), cultivated on the Ecuadoran coastal plain since 1953 and in the Amazonian lowland since 1979 (Carrión and Cuvi, 1985). It also omits such relatively new introductions as *Chrysalidocarpus lutescens, Phoenix canariensis, Roystonea regia, Trachycarpus fortunei,* and *Washingtonia robusta*—all ornamentals—and two probably ancient introductions, cultivated for their fruits: *Aiphanes aculeata* (most likely introduced from Colombia and grown on the coastal plains) and *Cocos nucifera* (commonly cultivated on the coastal plains and occasionally in the eastern lowlands).

APHANDRA

A monotypic genus, *Aphandra* has *A. natalia* as its only species. Despite its many uses and economic importance, *A. natalia* was not described until 1987 (Balslev and Henderson) in the genus *Ammandra*. It was later transferred to *Aphandra* (Barford, 1991a).

Aphandra natalia is a subcanopy palm, its trunk seldom growing taller than five meters. It is widespread on terra firma (unflooded forest) in Ecuador's southeastern lowlands, occurring up to an altitude of 800 meters. Probably endemic to northern Peru and southeastern Ecuador, it ranks among Ecuador's most economically significant palms. Most of the nation's brooms are made from the petiole or leafbase fibers of this palm. The fibers are extracted primarily from wild individuals, although limited cultivation of *A. natalia* has been observed among both colonists and Quichua and Achuar Indians. In addition, the edible fruits (mesocarp) of the palm are sold at the Sunday market in Sucua (Province of Morana-Santiago).

Various forms of in situ management are used to foster the growth of *A. natalia*. Quichua Indians of Canelos cut back other surrounding vegetation, maintaining that it promotes the quality and quantity of fibers. Around Huambi, south of Sucua, the palm is often left standing when the forest is cleared for pastures. Thus, we found that a specific plot of pasture land measuring only one-tenth of a hectare supported 19 palms, inferring that the per-hectare carrying capacity is 190 palms. There appeared to be no obvious signs of competition between palms and grasses. According to the land's owner, the income from selling fibers is several times that derived from raising cattle on the same plot.

Another type of management allows the spontaneous regrowth of *A. natalia* on land where vegetation is selectively cut and grazing is eliminated. In such conditions, we observed that a one-tenth-hectare plot supported 56 palms from which fibers had been harvested at least once and 68 juvenile palms not previously harvested. Thirty-nine of the juveniles would probably be ready for harvesting within two years' time (Borgtoft Pedersen and Balslev, 1990). In addition, palm seedlings were numerous beneath the shade of the stand, indicating the site's excellent potential as a nursery.

In harvesting, the palm leaves are cut with a machete, and the fibers are removed from the leaves by hand. About six leaves are typically left to secure the palm's growth. Once cleaned and cut to proper length, the fibers are sold to local middle men or directly to the broom factories.

Noncommercial uses of *A. natalia* also have been observed. Shuar, Achuar, and Quichua Indians occasionally use the leaves for thatch, though it is not considered of high quality. Blowgun darts are carved from the leaf rachis, and the palm's young, unopened spear leaves are used to make sheaths to protect the darts once they have been dipped in curare. Also, the palm heart is edible, as is the immature liquid or gelatinous endosperm. Because rodents also favor the fruits, areas supporting many *A. natalia* are considered excellent hunting grounds.

ASTROCARYUM

Four species of *Astrocaryum* have been recorded in Ecuador, of which *A. chambira* is the most commercially valuable. Medium to tall in size with a spiny trunk, this palm grows on terra firma in the eastern lowlands. The palm's hearts and fruits are edible. However, it is primarily exploited throughout Amazonian Ecuador for its fibers, which are used to make hammocks, fishing nets, string carrying bags, and other items. The string bags (or *chigras*) and hammocks are in great demand with tourists and are sold in many shops. Although the overall economic importance of these specific products is low, their sale often provides one of the few sources of cash for the indigenous groups producing them.

The fibers used to make these products come from the unopened spear leaves of the palm. How they are harvested depends on the palm's size. Because the spiny trunks are difficult to climb, tall palms are cut down. In shorter individuals, all the palm leaves may be cut until the spear leaf can be reached with a machete. In young, very short-trunked palms, only the spear leaf itself is removed. Once harvested, the fibers are extracted from the leaf's pinnae by stripping the upper layer and leaving the fibrous part. The fibers are then boiled and dried and, afterward, twisted into strings by rolling them back and forth on the thigh.

According to Davis and Yost (1983), *A. chambira* can be overexploited if a village of Indians, such as the Waorani, remains in one place for a long time. And tourists' growing demand for hammocks and string bags is likely to increase the palm's scarcity throughout the region. To ensure the palm's economic and cultural importance, it will probably have to be cultivated or harvested in nondestructive ways.

Minimal management of the palm has been observed. *A. chambira* was found growing near the houses in the Achuar village of Mashient. Though not actually cultivated, the young palms had sprouted from seeds that had been gathered in the forest and deliberately scattered in the gardens. Because most of the new leaves had been harvested, the palms did not look well. Nevertheless, they were productive.

Found on Ecuador's coastal plains, *Astrocaryum stanleyanum* is of minor economic importance. Both the fruits and palm heart are edible, and the pinnae taken from the palm's spear leaf are used to make hats sold in the province of Esmeraldas and elsewhere.

Astrocaryum jauari grows in dense stands in Amazonian Ecuador, in areas periodically inundated by lakes and black-water rivers. Locally, the Siona Indians use the fruit as fishbait and the hard endocarp—which separates the flesh from the endosperm, or seed—to make necklaces (Balslev and Barford, 1987). Siona children snack on the oily seeds. In Brazil, *A. jauari* seeds were once the basis for a small oil industry at Barcelos on the Rio Negro (Lleras and Coradin, 1988). Currently, the palm is being commercially exploited there, this time for its palm heart—a business that may seriously deplete large stands of the species.

Ecuador's fourth *Astrocaryum,* like *A. jauari,* is not commercially exploited in the country. The leaves of *A. murumuru* are used in the eastern lowlands during Easter processions, and Quichua Indians drink the liquid endosperm of the palm's immature fruits. In Para, Brazil, a small industry derives oil from the seeds (Balick, 1985).

ATTALEA

The only species of *Attalea* found in Ecuador, *A. colenda,* is endemic to the coastal plains of Ecuador and southern Colombia (Balslev and Blicher-Mathiesen, 1991). It is a huge canopy palm found in the perhumid forest, as well as in the drier areas of southern Ecuador.

Attalea colenda is commercially exploited for its oily seeds. This industry is based in the town of Manta. After the fruits are collected where they fall, the edible apricot-flavored mesocarp is removed, and the hard endocarp is cracked between two stones or with a hammer. (In the province of Esmeraldas, the mesocarp is eaten fresh or sun-dried.) The seeds inside are then sold to local middlemen.

Normally, this palm species is left intact when forested lands are cleared for agriculture, and it is common in many pastures. However, a study by Balslev and Blicher-Mathiesen (1991) demonstrates that the palm does not regenerate in open pastures. Because so little forest remains on Ecuador's coastal plains and because *A. colenda*'s distribution is very limited, Balslev and Blicher-Mathiesen conclude that the palm is endangered. Their study also shows that the quantity and quality of oil derived from *A. colenda,* in its present wild state, may equal that obtained at the beginning of the century from the African palm (*Elaeis guineensis*). This suggests that *A. colenda* could be cultivated to obtain its marketable oil.

BACTRIS

The genus *Bactris* (currently being revised by B. Bergmann, AAU) probably includes 15 species that occur in Ecuador (Balslev and Barford, 1987). The best known species is *B. gasipaes,* a usually spiny palm that also is cespitose, meaning that it grows in clusters and, thus, has numerous stems. Domesticated thousands of years ago, the palm is now grown throughout the Neotropics and is known only in cultivation.

Its fruits are often sold in lowland markets, either raw or prepared, but never—as in Colombia—in mountain markets. Canned fruits, common in Colombia, do not exist in Ecuador. The first palm-heart plantation for *B. gasipaes* was recently established in Amazonian Ecuador. The species has long been cultivated for this purpose in Costa Rica and Colombia. Because *B. gasipaes* is cespitose, its hearts can be harvested without killing the palm. According to Clement and Mora Urpi (1987), the species also has great potential as a source of oil and starch.

It is interesting to note that a germ plasm plantation containing 400 varieties of *B. gasipaes* from Ecuador, Peru, and Colombia has been established at INIAP (Instituto Nacional de Investigaciones Agropecuarias) near Coca in Ecuador's eastern lowlands. Studies there should enhance the productivity of the species.

To the indigenous people of Ecuador's lowlands, *B. gasipaes* is one of the most important crop plants. Its fruits are used to prepare *chicha,* a fermented beverage, or they are eaten after being boiled or roasted. The Indians recognize and cultivate many varieties of *B. gasipaes*—some that yield fruits high or low in oil, high or low in starch, with or without seeds, with spiny stems or smooth stems, and so on. Fruits are harvested from spiny varieties with a long, hooked pole or by climbing an adjacent *Cecropia* tree, which Quichua and Waorani Indians plant next to the palms. In addition, the palm's hard black wood is used to make spears and blowguns, artifacts that are now produced in large quantities for the tourist trade.

Many other species of *Bactris* bear edible fruits. The fruits of one are occasionally sold on the coastal plains (B. Bergmann, pers. comm.).

CEROXYLON

The genus *Ceroxylon* probably includes eight species—commonly called wax palms—found in Ecuador (Galeano, pers. comm.). The palms are usually tall and grow in the mountains between altitudes of 1,000 and

3,000 meters. The trunks are white (owing to a wax covering), smooth, and bear conspicuous internodes. Acosta Solis (1961) described the economic value of a *Ceroxylon* species called tamban. A number of products were obtained from this palm, such as wax, leaves for thatch, and "spear leaves," which are sold at Easter. Pinnae or leaflets taken from the spear leaves also were used for weaving mats, hats, and belts. Once used to make candles and other items, the wax currently has no known economic use in Ecuador, although the leaves are still sold in large quantities for Palm Sunday. Also, small baskets or crosses, which are then blessed by priests, are commonly made from the yellow, soft, and flexible pinnae. The trunks of tamban are used locally for fencing and to fashion pots for plants. And farmers around Cosanga, in the province of Napo, claim that the palm's fruits serve as excellent pig feed.

Ceroxylon often grows in dense stands, and—in certain parts of Ecuador—such stands are commonly found in pastures. However, these palms are in poor condition: they do not regenerate in pastures and yet continue to be cut down each year during harvesting of the leaves in the Easter season. Conditions for growth in pastures are not optimal either. As a result, in stands observed in pastures around Cosanga, the stems of most individual palms are so narrow just below the leaves that the entire crown often breaks off. This factor also accelerates depletion of the palms. On the other hand, *Ceroxylon* stands found in forested areas surrounding the pastures show vigorous regeneration; the forest floor is covered with seedlings (123 were counted in one square meter), and palms of all sizes are abundant.

Once young palms growing in the forest's shade start to form trunks, their spear leaves are as large or larger than those of adult individuals. In addition, these leaves can be harvested with little effort and cause little permanent damage to the palm. Regretably, farmers tend to cut the mature palms in their pastures because there are no nearby forests harboring young *Ceroxylon* palms—or at least not enough to supply the desired harvest.

Nevertheless, the strictly seasonal demand for *Ceroxylon* leaves bodes well for sustainable management, because it allows palms to regenerate between harvestings. Initial suggestions for management seem simple: keep the surrounding forest more or less intact and harvest leaves from some young individuals while leaving other young ones to mature and the older ones to bear fruit. Unfortunately, the forests that tend to support *Ceroxylon* are gravely threatened by deforestation, and current harvesting practices include cutting down adult palms. As the distribution of most of Ecuador's *Ceroxylon* species is limited, the entire genus must be considered endangered.

EUTERPE

Ecuador probably harbors five species of *Euterpe* (Balslev, 1990). Known from the coastal plain, *Euterpe chaunostachys* is one of two species (the other is *Prestoea trichoclada*) that are the main source of palm hearts for the canning industry. In addition, the palm's fruits and mesocarp are used to make ice cream and juice, which are marketed on a limited scale in the province of Esmeraldas.

A medium-sized cespitose palm, *E. chaunostachys* grows in almost monospecific (single-species) stands (*oligarchic forests* of Peters et al., 1989) on water-logged soils in estuaries of northwestern Ecuador. In Colombia, the palm is referred to as *E. cuatrecasana* and possibly is co-specific with (in other words, the same species as) *E. oleracea,* which is widespread in the Amazon region.

Although *E. chaunostachys* is occasionally cultivated around rural houses, only natural stands are exploited. We observed a one-tenth hectare plot in such a natural stand that supported 37 clusters (same as individuals), each having at least one stem that measured seven centimeters in diameter at breast height (more or less the minimum size needed for cutting the palm heart). The 37 clusters bore a total of 126 mature stems.

When the plot was harvested, 65 stems were cut, leaving 61. No mature stems were left in 30 percent of the clusters, while more than one was left in 43 percent (Borgtoft Pedersen and Balslev, 1990). The harvesters, who are paid according to the number of palm hearts they cut, were told by the company to leave at least one mature stem in each cluster to promote regeneration. Nevertheless, the number of stems left depended more on the palm's size, accessibility, and age (old trunks are hard to cut) than on company directions. As each of the 37 clusters bore many young shoots (and there were numerous seedlings and juvenile plants as well), there is little doubt that the stand will recuperate. The sustainability of this type of management depends very much on the schedule: Will the palms be allowed to regenerate before the next harvest?

Still wrapped in some of the protective leaf sheaths (which keep the palm hearts fresh for up to two days), the harvested hearts are taken to the factory. There, the leaf sheaths are removed, and the palm hearts are cut into proper lengths, washed, and then canned in water with citric acid. After the sealed cans are boiled, they are labeled and ready for national and international markets. Two canning factories are operating in the province of Esmeraldas, one near Borbon, and one in Atacames.

Another *Euterpe* species, the single-trunked *E. precatoria,* occurs in the Ecuadoran Amazon. It has no known commercial uses, though its heart is eaten throughout the region. Quichua Indians boil the roots to prepare a concoction to wash their hair. Also, the leaves are occasionally used for thatch, but are not considered of high quality.

IRIARTEA

According to Henderson (1990), *I. deltoidea* is the only species in the genus *Iriartea.* It is found in the lowlands on both sides of the Andes, growing up to around 1,000 meters, and may be Ecuador's most common palm (Balslev and Barford, 1987). Extremely hard and heavy, the trunks of *Iriartea deltoidea* are cut into poles and sold to banana plantations on the coastal plains. The poles are used to support the banana plants. Small industries, as in the town of Puyo, recently started using the wood to create furniture that is both durable and attractive.

I. deltoidea is very important to indigenous people. Its trunks, which are split to form boards, are used to build the walls in traditional Shuar and Achuar Indian houses. The houses' poles and floors also are made from the wood, as are blowguns. The palm's edible heart is considered very tasty, and Achuar and Quichua Indians also eat the palm's seed when still young and soft.

Iriartea deltoidea is often left standing when the forest is cleared, although cultivation or management of the palms has not been observed. It does not regenerate outside the forest.

JESSENIA

According to Balick's revision (1986), *Jessenia* is a monotypic genus; its sole species is *J. bataua.* As with *Iriartea,* this palm is found on both sides of the Andes and typically occurs up to an altitude of about 1,000 meters. (At one location, it was found growing at 1,350 meters.) A large canopy palm, it is often reported to be quite common on water-logged soils (Kahn, 1988), but in Ecuador, *J. bataua* has been observed to exist almost entirely in small groups or as scattered individuals on firm soils.

Although the palm is often mentioned as having great potential as an oil crop (Balick, 1985), it has little known economic importance in Ecuador. Oil from its mesocarp can be bought in towns in the eastern lowlands and in health food stores in Quito. It is usually extracted simply by mashing and boiling the fruits and skimming the oil from the surface. Its principal use is as a hair oil to prevent dandruff and hair loss and to improve hair luster and growth. Almost identical to olive oil, it also is good for cooking, but its price ($6.5 per liter in the lowlands, far more in Quito) is prohibitive.

The seeds of *J. bataua* occasionally are sold at markets to make potions for preventing hair loss. Also, the fruits are sold as a type of snack (only the oily flesh is eaten) at several lowland markets, such as in Coca and Borbon.

The fruit production of *J. bataua* has been briefly studied. For example, we observed nine palms during a period of 316 to 370 days and found that they produced 1.1 to 4.9 flowers per year (Borgtoft Pedersen and Balslev, 1990). In an earlier study, it was found that the weight of the fruit resulting from each *J. bataua* bloom may vary from five to 25 kilograms (Balick, 1981; Markley, 1949). It also was observed that if the mesocarp makes up 40 percent of the fruit and contains between 12.4 and 18.2 percent oil (Pesce, 1985), the resulting production will be 0.3 to 8.4 kilograms of oil from each palm per year.

It was further observed that palms growing in the forest's shade apparently begin to produce fruit at a much later stage than do palms exposed to sunlight. On shaded palms, flowers have not been noticed to appear below six meters, while on palms receiving sunlight, blossoms occur below three meters, and one was found growing only 1.2 meters above the ground. According to Sist and Puig (1987), *J. bataua* requires shade for germination and initial growth, but needs sunlight to start vertical growth.

Highly valued among most indigenous groups, the fruits of *J. bataua* are edible and are used to produce *chicha* and oil. The palm heart also is considered a delicacy, as are the larvae (*Rhyncophorus palmarum*) that are collected from the uppermost part of a palm's decaying trunk. Cayapa and Kwaiker Indians of the coastal plain use the palm's leafbase fibers to make blowgun darts (Barford and Balslev, 1988).

Actual management of *J. bataua* has not been observed, although we do know that the fruits are normally harvested by cutting down the entire palm. Felling also makes it easier to reach the palm heart and, eventually, the larvae. (The larvae will be ready to eat two to three months later.) However, around settlements such as the Quichua village of Canelos and the Achuar village, Mashient, the palms often are extremely rare because they are felled by both colonists and Indians.

PARAJUBEA

The genus *Parajubea* includes two species: *P. cocoides,* which is cultivated in Ecuador, and *P. torallyi,* which occurs in the wilds of Bolivia. According to Moraes and Henderson (1990), *P. cocoides* is probably a cultivated form of *P. torallyi,* yet is sufficiently different to warrant its acceptance as a separate species.

Cultivated throughout the Ecuadorean mountains up to an altitude of about 3,000 meters, *Parajubea cocoides* is so common in Quito that it is called the Quito palm. It is grown primarily as an ornamental, although the edible seeds can be bought at highland markets and are collected from below palms growing in parks and along streets. Occasionally, small trinkets are made from the fruit's hard endocarp and sold in shops.

PHYTELEPHAS

Two species of *Phytelephas* are found in Ecuador: *P. aequatorialis* and *P. macrocarpa.* The former is endemic to Ecuador's coastal plains; the latter is distributed in the western Amazon, from Colombia to Bolivia (Barford, 1991a).

A medium-sized palm reaching a height of 15 meters, *P. aequatorialis* occurs from sea level up to 1,500 meters. It also is commonly found in disturbed areas, such as steep slopes and floodplains. This palm has a great many uses. Both colonists and indigenous people use the leaves for thatch. The fruit's mesocarp is edible and is used to feed pigs and hens, to attract rodents when hunting, and to bait both fish and rodent traps. (The earlier population in the province of Esmeraldas used to extract oil from the mesocarp by boiling the fruit and skimming the oil from the water's surface.) In addition, the liquid endosperm is drunk; when it's older and jelly-like, it is eaten.

The haustorium of germinating seeds from *P. aequatorialis* can also be eaten and has a coconut-like flavor. At one time, the palm's leaf-sheath fibers were used for brooms, but the sheaths are most difficult to extract, and the practice is now rare. Rope has been made from the fibers as well, although—according to a farmer from the Alluriquin area—only fibers from male plants were appropriate. The palm heart, too, is tasty but hard to remove.

It is the mature seeds (or vegetable ivory) of *P. aequatorialis* that are most economically significant. As mentioned earlier, the international demand for products made from vegetable ivory, especially buttons, has returned and is increasing rapidly. It also is used to make other products in Ecuador, such as jewelry, chess pieces, and toys, and to inlay or decorate furniture. Manta is the main production center for buttons, while jewelry is primarily made in Riobamba, Quito, Guayaquil, and Salinas. Some leftovers from button production are milled and used, mixed with cotton press cake, as animal fodder.

Even though the demand for vegetable ivory had disappeared, farmers found that *P. aequatorialis* provided enough other local products so that it was worthwhile to maintain at least some of the palms in their agricultural systems. Thus, *Phytelephas aequatorialis* is often the only woody plant found intact in pastures, and, in the Rio Santiago area, it is a common component in agroforestry systems.

Managing the species may involve cutting down some of the male palms, as well as harvesting leaves from males rather than from the fruit-producing females. Around Alluriquin, three to four leaves are commonly left to ensure the palm's survival. Preliminary data from two study plots in the area suggest that male palms produce more leaves than do females, and that the rate of leaf production is slower in palms growing in the shade than among those receiving sunlight. At the same time, leaves from shaded palms appear to be longer, healthier, and greener, while fruit production is apparently greater in palms exposed to sunlight. The average number of infructescences on shaded palms is 7, while the average number on sunlit palms is 13.

The other species of *Phytelephas* is typically found on floodplains, sometimes forming large stands on periodically inundated soils in Amazonian Ecuador. The fruit mesocarp, immature endosperm, and the palm heart of *P. macrocarpa* are eaten. The palm's vegetable ivory is currently not extracted, probably because its seeds are much smaller than those of *P. aequatorialis*. Because the palm is low, has a thin, easy-to-cut stem, and grows in large stands, Quichua and Achuar Indians often prefer its leaves for thatch. Gathering leaves from more dispersed, harder to harvest palms is not an easy task: as many as 5,000 leaves are needed to thatch one Achuar house. In the upper Rio Napo region, the leaves are occasionally sold for use as thatch (M. Mamallacta, pers. comm.).

PRESTOEA

Probably six of the species belonging to the genus *Prestoea* are found in Ecuador (Balslev and Barford, 1987). Growing on the western Andean slopes at altitudes between 1,000 and 2,000 meters, *P. trichoclada* is—

along with *Euterpe chaunostachys*—Ecuador's commercial source of palm hearts. It is a medium-sized cespitose palm, as is *E. chaunostachys*. Thus, both sprout again after cutting. In some areas, *P. trichoclada* is very common, though it does not form stands as dense as those of *E. chaunostachys*.

The palm hearts are canned in Quito. (Quito supermarkets also occasionally offer fresh palm hearts, still enclosed in the crownshaft.) Palm-heart harvesters are not employed by the canning factories, but are paid according to the number of palm hearts they can deliver when the factory trucks arrive. No management practices have been observed, nor do any studies of the ecology of *P. trichoclada* exist. If it can be grown in the open, rather than in the forest, this palm may be an economically sound choice for reforesting large numbers of degraded pastures.

DISCUSSION

Palms are one of Ecuador's most important plant families for providing a variety of benefits and products. Certainly, the array of palm products with commercial value is extraordinary.

Although the extraction of perishable products from palms growing in remote areas is not feasible because of transportation problems, vegetable ivory and different types of fiber are among the most commercially significant palm products. Both can be stored until transportation is available, or until enough of the ivory or fiber has been harvested to constitute a shipment. Oil-bearing seeds also can be stored but fruits having mesocarp oil cannot as the latter deteriorates rapidly once it matures. This probably explains why the extraction of oil from the mesocarp of *Jessenia bataua* fruits is still a cottage industry, while removing oil from the seeds of *Attalea colenda* is a full-fledged business. Palm hearts also are highly perishable; however, because the palm hearts do not mature at a specific time as do fruits, they can be harvested at any time throughout the year. Therefore harvesting can wait until transportation is available and storage of the palm heart can be avoided.

Furthermore, palms are an ideal harvestable crop in areas unsuited or little used for agriculture, such as lands that are periodically inundated or water-logged. Why? Because there they tend to form oligarchic forests (forests dominated by one species), or they at least occur in high densities. In Ecuador, this is true of *Euterpe chaunostachys*, *Phytelephas aequatorialis,* and *Phytelephas macrocarpa,* as well as of some species not currently exploited commercially, such as *Mauritia flexuosa* and *Astrocar-*

yum jauari. But palms also are common on terra firma: Balslev et al. (1987) found that in a one-hectare plot near Anango, in the eastern Andes lowlands, 13 percent of the individual trees having a diameter at breast height of more than 10 centimeters were *Iriartea deltoidea*. And, as mentioned earlier, wax palms (*Ceroxylon sp.*) may form large stands on the slopes of the Andes.

Finally, palms offer other benefits, not all of which are of commercial value, and are often left in pastures and agricultural areas when forests are cleared so as to retain these benefits. For example, cattle farmers around Sucua report that *Aphandra natalia* was originally left in pastures because its fibers were used for making rope to tie up cattle. Although this use was later abandoned when inexpensive synthetic rope became available, it ensured the palm's survival until the broom industry began to use *A. natalia* fibers. Similarly, *Phytelephas aequatorialis* was not cut down, despite the lack of a market for vegetable ivory, because the palm and its products had so many other uses. Palms also help reduce erosion, are more resistant to fires than are dicotyledons, are valued as ornamentals, and provide shadow for cattle in pastures (albeit a small amount). One last point: Palms left in agricultural areas are far easier to harvest than are those found in remote, natural forests.

PALM HARVESTING AND CONSERVATION

Commercial harvesting may be an important tool in saving portions of the tropical forests from agricultural conversion, because it provides an economic incentive for maintaining the forests. Furthermore, commercial harvesting may also lead to cultivation and domestication of new crop species, thereby diversifying the agricultural systems. However, one must be aware of the limitations of this form of land use. Harvesting rain forest products on a commercial scale will inevitably affect the forests in one way or the other; often the impact will be significant. For example, when products are extracted, nutrients are removed. Of even greater concern is the impact of destructive harvesting, such as felling palms or trees to obtain their products or favoring desired species by cutting down others.

Regretably, destructive harvesting is exacerbated by a lack of adequate harvesting tools, methods, and knowledge of forest species. In addition, harvesting is often done on lands having no recognized ownership, which may result in several companies or individuals exploiting resources at the same time. In such cases, overharvesting usually occurs as well because everyone is hoping to beat the competition. Strict controls seem essential to reduce injurious exploitation. But enforcing such controls in remote regions, where most harvesting occurs, will be extremely difficult.

Examples of destructive harvesting abound: in Brazil, *Euterpe edulis* has been eliminated in many areas because of overharvesting for the palm hearts (Nodari and Guerra, 1986). In Chile, the Chilean wine palm (*Jubaea chilensis*) is now listed as endangered by IUCN (Johnson, 1988), in part because it is being felled to obtain sugar sap from the trunk (Corner, 1966; Moore, 1979). In Peru, around Iquitos, *Mauritia flexuosa* stands have been seriously depleted because of detrimental harvesting practices used to supply the city's markets with the palm's fruits (Padoch, 1988). As previously described, *Jessenia bataua* is often scarce around permanent settlements in Ecuador because it is cut down to obtain the fruit. Already, the commercial harvesting of *Astrocaryum chambira* in Ecuador is so intensive that the palm is now rare in certain areas. When current commercial exploitation increases, as it most likely will, the impact on most palm species will intensify.

Although destructive harvesting is probably one of the greatest problems in the commercial exploitation of palms, the favoring of economically valuable species is a significant concern as well. Around the village of Canelos, other vegetation found around *Aphandra natalia* is cut down so as to promote the palm's growth. Also, because both *Phytelephas aequatorialis* and *Jessenia bataua* are more productive when exposed to sunlight, surrounding vegetation is removed. These three species probably are but a few examples of the palm species—and tropical forest plants in general—that will benefit, at least in the short term, from a more sun-filled environment. Selectively cutting plants of no commercial value is likely to become common practice in harvesting those that are commercially significant. Thus, even if destructive harvesting practices are kept to a minimum, favoring the growth and production of commercially valuable species will have a detrimental effect on forests as a whole.

A closing thought: Even if the commercial exploitation of tropical forest species results in forests being converted to semi-natural systems, in which wild and cultivated plant species mix, such systems will still be more biologically diverse than pastures or agricultural lands are. In addition, they will serve as gene banks. Of course, there must always be a clear distinction between lands earmarked as extractive reserves and those that exist as national parks and protected areas. Commercial exploitation can only benefit the tropical forests if it replaces modern agricultural practices, not if it replaces essential protected natural areas.

References

Acosta Solis, M. 1944. *La Tagua*. Editorial Ecuador, Quito.

Acosta Solis, M. 1961. *Los bosques del Ecuador y sus productos*. Editorial Ecuador. Quito.

Balick, M. J. 1981. *Jessenia bataua* and *Oencarpus* Species: Native Amazonian Palms as New Sources of Edible Oil. In E. H. Pryde, L. H. Pincen, and K. D. Mukherjee, eds., *New Sources of Fats and Oil*. Champaign, IL: American Oil Chemists Society.

Balick, M. J. 1985. Current State of Amazonian Oil Palms. In C. Pesce, *Oil Palms and Other Oilseeds of the Amazon*. Algonac, Michigan: Reference Publications, Inc.

Balick, M. J. 1986. Systematics and Economic Botany of the *Oencarpus-Jessenia* (Palmae) Complex. *Adv. Econ. Bot.* 3:1–140.

Balslev, H. 1988. Distribution Patterns of Ecuadorean Plant Species. *Taxon* 37(3):567–77.

Balslev, H. 1990. Palms of Ecuador. In S. Laegaard and F. Borchsenius, (eds.), *Nordic Botanical Research in the Andes and Western Amazonia*. AAU Reports 25.

Balslev, H. J., Luteyn, B. Ollgaard, and L. B. Holm-Nielsen. 1987. Composition and Structure of Adjacent Unflooded and Floodplain Forest in Amazonian Ecuador. *Opera Bot.* 92:37–57.

Balslev, H., and A. Barford. 1987. Ecuadorean Palms: An Overview. *Opera Bot.* 92:17–35.

Balslev, H., and A. Henderson. 1987. A New *Amandra* (Palmae) from Ecuador. *Syst. Bot* 12:501–504.

Balslev, H., and U. Blicher-Mathiesen. 1991. La "palma real" de la Costa Ecuatoriana (Attalea colenda, Arecaceae) un recurso poco conocido de aceite vegetal. In M. Ríos and H. Borgtoft Pedersen, (eds.), *Las Plantas y el Hombre*. Herbario QCA and Abya-Yala. Quito.

Barford, A. 1989. The Rise and Fall of Vegetable Ivory. *Principes* 33:181–90.

Barford, A. S. 1991a. A Monographic Study of the Subfamily Phytelephantoidea (Arecaceae). *Opera Bot.* 105:1–73.

Barford, A. S. 1991b. Usos pasados, presentes y futuros de las palmas Phytelephantoidees (Arecaceae). In M. Ríos and H. Borgtoft Pedersen (eds.), *Las Plantas y el Hombre*. Herbario QCA and Abya-Yala. Quito.

Barford, A., and H. Balselev. 1988. The Use of Palms by the Cayapas and Coaiqueres on the Coastal Plains in Eucador. *Principes* 32:29–41.

Barford, A., B. Bergmann, and H. Borgtoft Pedersen. 1990. The Vegetable Ivory Industry: Surviving and Doing Well in Ecuador. *Econ. Bot.* 44(3):293–300.

Borgtoft Pedersen, H., and H. Balslev. 1990. *Ecuadorean Palms for Agroforestry*. AAU Reports 23.

Cañadas Cruz, L. 1983. *El mapa bioclimático y ecológico del Ecuador*. Banco Central del Ecuador. Quito.

Carrión, L., and Cuvi, M. 1985. *La palma africana en el Ecuador: Tecnologia y expanción empresarial*. Facultad Latinoamericano de Ciencias Sociales (FLASCO). Quito.

Clement, C. R., and J. Mora Urpi. 1987. Pejibaye Palm (*Bactris gasipaes,* Arecaceae): Multi-Use Potential of the Lowland Humid Tropics. *Econ. Bot.* 41:302–11.

Companía "Guia del Ecuador." 1909. *El Ecuador.* Guia commercial agricola e industrial de la Republica. Compañía "Guía del Ecuador." Guayaquil.

Corner, E. J. H. 1966. *The Natural History of Palms.* London: Weidenfeld and Nicolson.

Davis, E. W., and J. A. Yost. 1983. The Ethnobotany of the Waorani of Eastern Ecuador. *Bot. Mus. Leafl.* 29(3):159–211.

Descola, P. 1989. *La Selva Culta.* Simbolismo y praxis en la ecología de los Achuar. 2' Spanish edition. MLAL and Abya-Yala. Quito.

Galeano, G., and R. Bernald. 1987. *Palmas del Departamento de Antioquia.* Universidad Nacional de Colombia, Bogotá.

Henderson, A. 1990. *Arecaceae Part I. Introduction and the Iriarteinae.* Flora Neotropica.

Holdridge, L. R., W. C. Grenke, W. H. Hathway, and J. A. Tosi, Jr. 1971. *Forest Environments in Tropical Life Zones: A Pilot Study.* Oxford: Pergamon Press.

Johnson, D. V. 1988. Worldwide Endangerment of Useful Palms. *Adv. Econ. Bot.* 6:268–73.

Kahn, F. 1988. Ecology of Economically Important Palms in the Peruvian Amazon. *Adv. Econ. Bot.* 6:42–49.

Karsten, R. 1935. *The Head Hunters of the Western Amazon: The Life and Culture of the Jíbaro Indians of Eastern Ecuador and Peru.* Societas Scientiarum Fennica. Commentationes Humanarum Litterarum. VII.1. Helsingfors.

Lleras, E., and L. Coradin. 1988. Native Neotropical Oil Palms: State of the Art and Perspectives for Latin America. *Adv. Econ. Bot.* 6:201–13.

Markley, K. S. 1949. *FAO Oilseed Mission for Venezuela.* Washington, D.C.: FAO.

Moore, H. E., Jr. 1979. Endangerment of the Specific and Generic Level Inpalms. *Principes* 23:47–64.

Moraes R. M., and A. Henderson. 1990. *The Genus Parajubea (Palmae).* Brittonia 42.

Nodari, R. O., and M. P. Guerra. 1986. O palmiteiro no su. do Brazil: Situacao e perspectivas. *Useful Palms of the Tropical America* 2:9–10.

Padoch, C. 1988. Aguaje (*Mauritia flexuosa* L.f.) in the Economy of Iquitos, Peru. *Adv. Con. Bot.* 6:214–24.

Pesce, C. 1985. *Oil Palms and Other Oilseeds of the Amazon.* Rev. trans. by D. V. Johnson. Algonac, MI: Reference Publications, Inc.

Peters, C. M., M. J. Balick, F. Kahn, and A. B. Anderson. 1989. Oligarchic Forests of Economic Plants in Amazonia: Utilization and Conservation of an Important Tropical Resource. *Conservation Biology* 3(4):341–49.

Sist, P., and H. Puig. 1987. Régéneracion, dynamique des populations et dissémination d'un palmier de Guyane Francaise: Jessenia bataua (Mart.) Burret subsp. oligocarpa (Griseb. and H. Wendle) Balick. Bull,. Mus.natn.Hist.nat., Paris 4é sér., 9, section B, *Adansonia* 3:317–36.

Strotty, L., and G. Malagotty. 1950. *La agricultura en el Territorio Amazonas: Exploitación del seje (Jessenia bataua) palma oleaginosa.* Caracas.

Velasco, J. 1789. *Historia del Reino del Quito en la America Meridional. Historia natural. Casa de la Cultura* (1977). Quito.

21

The Jatata Project: The Pilot Experience of Chimane Empowerment

GUILLERMO RIOJA
Conservation International, Bolivia

The Maniqui River, which flows from south to north in the department of Beni, Bolivia, is one of the important tributaries of the Amazon Basin. Low tropical humid forest (mountainous humid forest of the eastern sub-Andean foothills) is found on the banks of this river and, to the east, this forest merges with the tropical humid forest of the lowlands. These two formations of vegetation are rich in palms. Of particular interest to us here is the jatata palm *Geonoma deversa,* commonly known in Bolivia as jatata.

This palm extends to Guatemala, Honduras, Niacaragua, Costa Rica, Panama, Colombia, Venezuela, and the Amazon Basin of Ecuador, Brazil, Bolivia, and Peru (Wessels Boer, 1989).

The Chimane [pronounced: chee-**mah**-neh] ethnic group in the department of Beni is settled from the headwaters of the Maniqui River to the network of plains situated north of the city of San Borja. Traditionally, the Chimane have participated in the regional economy by supplying products they extract from the forest. Among these Chimane products are fabrics made from the leaves of the jatata palm.

In spite of their long history of contact since colonial times with Bolivian society, the Chimane ethnic group has been able to retain remarkable social cohesion, unlike most indigenous cultures in lowland Bolivia. The Chimane also continue to struggle to win legal recognition for their territorial claims, which has now led to the creation of important endowments.

The group is, however, still involved in a range of highly inequitable trade relationships. In the specific case of the jatata fabrics, this trade clearly favors the river merchants who make exorbitant profits using a system of barter, the formula for which, C-M-C (in which C stands for "cash," and M for "merchandise") brings about high levels of compulsory exploitation. In order to understand this better, we must take into account that an adult Chimane needs two days to collect and braid seven jatata sheets. (The *Geonoma deversa* leaves are braided in lengths of from 1.8 to 2.0 meters.) The average daily wage in the region is 15 bolivianos, equivalent to about $5.00 (U. S. dollars). Thus, seven sheets manufactured over a two-day period should sell on the market for 4.28 bolivianos or $1.42 each. Nevertheless, the river merchants pay the tiny sum of from 0.05 to 0.40 bolivianos, before reselling the merchandise at 1.40 bolivianos or $0.46 at the market in San Borja. It must be noted that these asymetrical market relations between the river merchants and Chimanes translate into the systematic violation of the most elemental human rights.

We should also point out that, in terms of conservation, the commercial demand for jatata fabric has grown tremendously because of its superior quality as a roofing material, the good ventilation that it affords, its suitability for use with both rectangular and round roof beams, and its ability to repel against breeding insects. The Chimane have indicated that the jatata groves that can be harvested are disappearing from areas near their communities. Thus, they find they must dedicate ever more time to the collection of leaves and the manufacture of the fabric, so they no longer have time for traditional activities such as agriculture and hunting. Currently, Chimane families are leaving the area of the upper Maniqui for areas where there is no trade in jatata fabric.

This conflict-ridden situation can be divided under three headings: compulsory exploitation of Chimane labor, the random pillage of the resource, and the systematic violation of human rights. Endogenous solutions from the ethnic group itself have had to be developed that, as a first resort, have been promoted and made possible by the Jatata Project.

THE PROJECT

The "Research Project on the Management and Marketing of Jatata by the Chimane Natives of Eastern Bolivia" was proposed to Conservation International in January 1990 by the anthropologists Richard Piland and Guillermo Rioja. The latter redrafted and adapted the project to the actual regional conditions in the Beni. The project became known as the "Jatata Project."

The Jatata Project set forth the following objectives:

• Establishment of a Chimane Standing Commission, to be provided with training in the direct marketing of jatata fabric. (This group later took the name of "The Chimane Committee for Management");

• Identification of areas of prospective cooperation in the development of sustainable jatata production;

• Identification of non-Chimane groups and individuals who can play an important role in the marketing of jatata;

• Preliminary research on the demand for jatata and the market for it in San Borja;

• Marketing of jatata fabric by the Chimane Committee for Management.

By September 30, 1990, these objectives had been fulfilled and the Chimane Committee for Management had made direct sales of 1,500 jatata sheets, earning 600 bolivianos, equivalent to $200. These funds furnished the capital to continue the collection of jatata for subsequent sale. From that point on, the Jatata Project became self-sustaining, and direct financial support from Conservation International ceased.

ASSESSMENT

The Jatata Project demonstrated the appropriateness of a low-tech approach to development based on better marketing of traditional handicrafts. The project furnished an incentive for creative proposals from the Chimanes, who were catalyzed in this process by their organization, the "Tsimane' Council," which contributed through its cooperation to ethnic empowerment.

Furthermore, seemingly insurmountable obstacles were successfully confronted in a collective fashion. The case in point was that of the river merchant buyers of jatata fabric, often long-standing violators of the most fundamental human rights, who were unable to exploit the Chimanes once they had been organized.

The Jatata Project distilled group consciousness and ethnic identity; it also encouraged ethnic pride, the *sine qua non* for the qualitative leap that the group is making toward the multi-ethnic, poly-cultural consensus of a new Bolivian society.

The Jatata Project had an impact on academic research in the Department of Beni—in such fields as ethnobotany, social communication, and social-support anthropology—because it established a new way of perceiving the ethnic issue, a way that is conducive to policies encouraging ethnic empowerment. Along these lines, the Conservation Data

Center (CDC-Bolivia) is currently promoting a project that would provide a group of Chimanes training with a view toward the development of a proposal for jointly managed biological research in the zone. Moreover, the Beni Biological Station (EBB) began an Extension Program Project in November 1990 as part of its Educational Program on the Environment, Communication, and Public Use of the Biosphere Reserve. This project will provide training to 15 Chimanes for the purpose of forming a Chimane Communication Committee along the lines of the model implemented by the Jatata Project.

FOLLOW-UP

It should not be construed from the foregoing that the Jatata Project is a single self-contained endeavor. On the contrary, any project is the start or fulfillment of a process, whose implementation opens up other research and programmatic prospects that should be developed gradually, but without delay.

The following stage of the Jatata Project should consist of a profound study of the ecology of *Geonoma deversa* in order to propose scientific means of preservation and of ongoing sustainable resource utilization. Otherwise, the regional survival of this palm will be endangered. Therefore, a project has been proposed to Conservation International, called the "Jatata in the Eva-Eva Mountain Range: The Biology, Population Dynamics, and Ethnobotanical Study." It only awaits financing.

Moreover, socio-cultural study is needed to develop production and marketing options for alternative Chimane products for a more diverse utilization of forest resources. Such activity would prevent the Chimane from participating in the market as monoproducers of jatata fabric.

In this regard, the Chimane themselves have chosen to produce less jatata fabric, now that the price has risen and the pressure of demand may cause immediate adverse cultural impact, which merits consideration. The Bolivian Project of Conservation International is currently working on a coherent methodological proposal for this issue, and it should lead to an appropriate project that will involve the joint-management participation of the Chimane.

In conclusion, in its first stage, the Jatata Project has achieved success in shedding light on the Chimane ethnic issue and in promoting a viable alternative for the cultural survival of the group, all the while promoting appropriate use of the jatata resource in the region. Nevertheless, the process that will make possible the permanent management of nontimber resources of the forest by its ancestral inhabitants is far from over.

References

Baslev, H., and M. Moraes. 1989. *Sinopsis de las palmeras de Bolivia*, AAV Reports 20. Hervario Nacional de Bolivia.

Piland, R., and G. Rioja. 1990. *Proyecto de investigacion sobre el manejo y comercializacion de la* Geonoma deversa *por los indios chimane del Oriente de Bolivia.* CI-Bolivia.

Rioja, G. 1990. *Avances para la consecucion del Proyecto Jatata.* CI-Bolivia.

Rioja, G. and M. Moraes. 1990. *La Jatata en la serrania de Eva-Eva: biología, dinámica poblacional y estudio etnobotánico.* CI-Bolivia.

Wessels Boer, J. G. 1968. The Geonomid Palms. Verandelingen der Königninglijke Nederlandse Akademic van Wetenschappen. Atd. Naturkunde. 58:5–202.

22

The Chimane Experience in Selling Jatata

Jorge Añez
The Tsimane' Council, Bolivia

We, the Chimane, live on the Maniqui River. We number over 6,000, and we inhabit the Ballivian Province in the department of Beni, Bolivia.

Where we live, there is a palm that we call "cajtafa," known in Spanish as jatata. From this palm, we make roofing for our homes. Jatata is plentiful where we live. In a town close to where we live, called San Borja, the people buy our jatata, and almost all the houses use jatata. Other towns use it also.

Almost all the Chimane work the jatata because it is plentiful in our forest. The men, the women, and the children work the jatata. Working the jatata is difficult because first one must go and take it from the mountain. Then, at home, the whole family weaves it. On one trip to the mountain, enough jatata is gathered to make eight or ten sheets.

The producers of jatata earn very little by their work. This is because the merchants do not pay them a fair price. These merchants do not treat the Chimanes well when they come to trade with them for the jatata. At times, the merchants give alcohol to the Chimanes and then force them to take things they do not want. They exchange a bottle of liquor that costs four bolivianos (B$4.00) for twenty or thirty jatata sheets. Thus, the sheets that they make in two or three days would only cost, with their work, two bolivianos a day. In this way, they cheat them when they sell them other goods. Since the merchants force the products on them, the debt never ends. They give them a used shirt that costs five bolivianos for forty sheets of jatata. Very often after the Chimane has paid off his debt, the merchant forces him to continue paying by threatening him with a machete, revolver, or rifle, so the Chimane have to keep giving them jatata. This is how the merchants force the

Chimane to work against their will. Very often, when the Chimane refuse, the merchants fire their guns quite close to the Chimane, and in this way frighten them so they will keep working. Also with their guns, the merchants take advantage of the Chimane women. The merchants also take advantage of their children. The merchants do not want them to go to school, even though they do. The merchants use the Chimane as their servants, and that is why the Indians cannot care for their own farms or other things. Even though the merchants make the Chimane work very hard, the latter have nothing, not even their own clothing.

When the Chimane are wounded by the merchants, they go to the town to file complaints, but the authorities do not help.

The merchants take advantage of the Chimane because the Indians do not know the true worth of their jatata. Recently, Guillermo Rioja of Conservation International has visited us and has seen how the merchants abuse us when they buy our products. He has helped us to reduce the influence of the merchants, and he has trained us so we can sell our jatata sheets ourselves at a fair price.

He has given us a course on organizing a Chimane Committe for Management. Fifteen Chimanes participated in the course, and they make up the Committee. These young people are the ones who are convincing their relatives not to sell to the merchants. Guillermo Rioja has given us a small fund so that the Committee can begin to buy jatata to sell later. This way we can free ourselves of the merchants.

Since the Committee was created, the Chimane suffer less, and now there are also fewer merchants mistreating our relatives. This is because many Chimane no longer work for the merchants, and they go and sell their jatata to the Committee, which pays them a fair price. Now the Chimane sell fewer jatata sheets and earn more. Later the Committee sells jatata to those who wish to buy it. This is the way in which the Chimane can sell their jatata because, since they do not have a house in San Borja, they cannot sell their jatata there. Since the Committee lives there, it can store the jatata until someone comes to buy it. The Committee helps the Chimane live better.

The only problem that we have now is that we need more money to be able to buy more jatata from our relatives who are jatata producers. Our idea is that all of the Chimanes would be able to sell jatata to the Chimane Committee for Development, and in this way never again have to exchange their jatata with the merchants for things we do not need. But we do not want the jatata to disappear; we want technical help so that this does not come to pass.

IV

Plants as Medicines

23

Ethnopharmacognostic Study of Kuna Yala

MAHABIR P. GUPTA, ANA JONES, PABLO SOLÍS,
AND MIREYA D. CORREA A.
Universidad Nacional de Panama

The Comarca of San Blas covers an area of 3,206 square kilometers and a population of over 30,000 inhabitants. Even though hospitals, health care centers, and health posts exist in Kuna Yala, during the first census of 1982, this group showed a very high index (67 percent) of malnourishment. This constitutes the most important health problem of the region, followed by intestinal disorders, which principally affect infants less than one year old.

In the Cuna society, there are two important persons who are responsible for health care. They are "Nele," who diagnoses the diseases, and the "Inadulet," who prepares and administers remedies. They are equivalent in modern medicine to a physician and a pharmacist, respectively.

METHODOLOGY

The methodology employed in this study was similar to the one utilized by Joly and coworkers.[1] Ana Jones (one of the coauthors of this chapter) is of Cuna descent and comes from Ailigandi, in the San Blas Archipelago. She was a final year student in the College of Pharmacy of the University of Panama, and, as part of the requirements of obtaining her Licentiate degree in pharmacy, she carried out a special project in her home village. She was trained in the techniques of interviewing the locals and in the collection of plant specimens. Forms were specially designed so that she could document ethnobotanical uses, plant part(s), manner of preparation and administration of the remedies, adverse effects, and any other notes of interest. All the interviews were tape-recorded in Cuna and later translated into Spanish.

Two "inadulet" of Ailigandi, Efraín Tejada and Jimmy Johnston, were interviewed. They explained the principles of Cuna medicine, their different specialties, and the forms of preparation and administration of the remedies.

The pressed plant specimens were identified by Professor Mireya Correa, another one of the authors, and voucher specimens were deposited in the herbarium of the University of Panama.

RESULTS

Of 99 specimens collected, 90 were identified to genus and 78 to species. In all, 42 plant families were represented.

The Cuna therapy may be administered in three forms: 1) Curative enchantments—these play an important role in the Cuna medicine. The curative method is based on the utilization of anthropomorphic figures, which serve to counteract the bad "spirits," 2) therapy based on medicinal plants, and 3) therapy based on medicines prepared from bones.

The Cuna medicine is divided into:

• "*Ina-Binit*": This constitutes the collection of fresh plants when needed. It is divided into "inadualet"—meaning boiled medicine, indicating that the plants are boiled in water to make a decoction, and "*ina duar suilit*"—meaning unboiled medicine, indicating that the plants are macerated to prepare infusions for local baths.

• "*Ina sevret*": This is preserved medicine, a way of preserving the medicines so that they do not lose their potency.

• "*Ina Kuamakalet*": This type of medicine is prepared in a special mortar, in which the plant material is mixed with the excrement of red ants and moistened with water. Later, oval pills are prepared and dried in the sun. At the time of using the remedy, the pills are scratched with a stone and a little water and the resulting solution is used.

• "*Inadibialet*": Fresh leaves of the plant are placed in fire and the resulting ash is mixed with a colorant "Magep" and used topically.

The following list enumerates the plants identified in this study, in alphabetic order, with respect to families and to genera within the families.

V: Vernacular name. The vernacular names are those given by the "Neles" and "Inadulets."
MED: Plant part used, preparation, and ethnomedical use.
VS: Voucher Specimen.
PMA: Herbarium of the University of Panama.

ACANTHACEAE
Aphelandra hartwegiana. V: Arbon'gid. MED: A flower infusion is used rectally as a local bath for constipation. VS: Florpan No. 225 (PMA).

Justicia ephemera. V: Suegia dup. MED: The whole plant in the form of "Inadibialet" is placed as a cross on the chest and the arms to treat convulsions during sleep in children. VS: Florpan No. 222 (PMA).

Justicia sp. V: Uisin bunnu'gid. MED: Infusion of the whole plant is used for internal fever in weak children, to whom the "nele" has diagnosed death. VS: Florpan No. 276 (PMA).

AMARANTHACEAE
Amaranthus sp. V: Ikui Kinnid. MED: An infusion of the leaves and flowers is used as a body bath to treat the disease called "Poni kinnid," which is characterized by vomiting and blood in the urine. VS: Florpan No. 302 (PMA).

AMARYLLIDIACEAE
Crinum darienensis V: O'INA MED: A whole plant infusion is used as a bath for internal fever in weak children. VS: Florpan No. 271 (PMA).

APOCYNACEAE
Rhabdadenia biflora. V: Ailisir e'giar. MED: An infusion of the flowers is used for washing the eyes in cases of infection. VS: Florpan No. 299 (PMA).

ARACEAE
Anthurium penthaphyllum. V: Nali Oskon'gid. MED: A whole plant decoction is used as a warm bath for skin infections VS: Florpan No. 259.

Anthurium sp. V: Uer'uer guabin'gid dummat. MED. An infusion of the flowers prepared as in "Ina Kuamakalet" is drunk to reduce the size of the uterus in a pregnant woman. VS: Florpan No. 240 (PMA).

Dieffenbachia aurantiaca. V: Abiormachi. MED: A whole plant decoction is used as a body bath for aches and skin infections. This plant is alleged to be a skin irritant. VS: Florpan No. 275 (PMA).

Dracontium dressleri. V: Igar, Kausis sapi, Naibe naba sannuar. MED:

The ground fruit in the form of "Ina Kuamakalet" is drunk to facilitate childbirth, while the leaves and roots are taken for muscle aches and snakebites. VS: Florpan No. 237 (PMA).

Montrichardia arborescens. V: Puppur. MED: A decoction of the leaves or the medication prepared in the form of "Inadibialet" is used for skin infections. VS: Florpan No. 297 (PMA).

Philodendron sp. V: Abior tubaled. MED: A whole plant infusion is used as a body bath to increase vitality in children. The sap is placed on the wound to accelerate healing. VS: Florpan No. 264 (PMA).

Spathiphyllum quindiuense. V: Ina'ulu puruiguad. MED: A whole plant infusion is used as a bath for weak children, and in cases of children to whom the "nele" has diagnosed death, it is administered four times a day for eight days. VS: Florpan No. 272 (PMA).

Spathiphyllum friedrichsthalii. V: Ina'ulu. MED: A flower infusion prepared as "Ina Kuamakalet" is drunk to avoid the enlargement of the uterus in a pregnant woman. To facilitate childbirth, it is drunk from the sixth until the eighth month of pregnancy. VS: Florpan No. 236 (PMA).

Xanthosoma helleborifolium. V: Naibe'uar, Uer'Uer, Guabingid bipinguad. MED: The tubers prepared as "Ina Kuamakalet" remedy is drunk to facilitate childbirth. VS: Florpan No. 239 (PMA).

Xanthosoma mexicanum. V: Naibe'mor dup sipuguad. MED: A fruit infusion prepared as "Ina Kuamakalet" is drunk to facilitate childbirth and for snakebites. VS: Florpan No. 248 (PMA).

Xanthosoma sp. V: Kuidar. MED: A decoction from the inflorescence is drunk for "Kala Luki," which is a disease that causes strong muscular aches and bone aches and inflammation, leading to a total paralysis. VS: Florpan No. 247 (PMA).

BIGNONIACEAE

Arrabidaea verrucosa. V: Divrsabina dup. MED: An infusion of the stems is drunk for internal fever weakness and muscle aches. VS: Florpan No. 208 (PMA).

Stizophyllum riparium. V: Kuanur Dup. MED: A stem decoction is drunk or used as a local bath to dilate atrophic urethra in cases of anuria. VS: Florpan No. 207 (PMA).

CACTACEAE

Pereskia bleo. V: Butarrar. MED: A decoction of the leaves is used as a warm bath to treat muscle aches, while a remedy prepared from the inflorescence as "Ina Kuamakalet" is drunk for stomach aches. VS: Florpan No. 210 (PMA).

CAPPARACEAE

Cleome serrata. V: Uerki dup. MED: A whole plant decoction is used locally as a warm bath. A remedy prepared from the young leaves in the form of "Ina Kuamakalet" is drunk to treat snakebites. VS: Florpan No. 220 (PMA).

CARICACEAE

Carica papaya. V: Kuat Kuat. MED: A remedy prepared from flowers and leaves in the form of "Ina Kuamakalet" is drunk to improve blood circulation, to cause diuresis, and to treat furuncles. VS: Florpan No. 274 (PMA).

COCHLOSPERMACEAE

Cochlospermum vitifolium. V: Monor. MED: An infusion prepared from its cotton seed is drunk, or taken as "Ina Kuamakalet" to improve blood circulation. VS: Florpan No. 227 (PMA).

COMMELINACEAE

Commelina diffusa. V: Biski. MED: A whole plant infusion is used as a bath for common colds. VS: Florpan No. 285 (PMA).

COMPOSITAE

Neurolaena lobata. V: Ina Kaibit, Ina kinkub'gid. MED: A whole plant decoction is used as a cataplasm to treat skin infections. VS: Florpan No. 214 (PMA).

Rolanda fructicosa. V: Sapur butarrar. MED: A whole plant decoction is used as a local bath for muscle aches. It is said that the plant with spines or with star-shaped fruits is used for muscle aches. VS: Florpan No. 295 (PMA).

Vernonia sp. V: Uu'sipu. MED: An infusion of the inflorescence is used as a bath to facilitate expulsion of placenta. VS: Florpan No. 294.

CUCURBITACEAE

Cucurbita pepo. V: Moe. MED: A whole plant infusion is used as a bath for children to accelerate their growth. VS: Florpan No. 280 (PMA).

CYATHEACEAE

Cyathea petiolata. V: Tryon Viseb'le. MED: A whole plant infusion is used as a bath in a "cayuco" for weak children, to whom the "nele" has diagnosed death. This treatment is done in a special ceremony. VS: Florpan No. 261 (PMA).

CYCADACEAE

Zamia pseudoparasitica. V: Obser. MED: A decoction made from ground tubers is drunk to induce vomiting, while the stem is placed on the affected parts for muscle aches. VS: Florpan No. 280 (PMA).

CYCLANTHACEAE

Carludovica palmata. V: Naivar. MED: An infusion of the inflorescence prepared as "Ina Kuamakalet" is drunk by pregnant women to prevent complications during childbirth. VS: Florpan No. 238 (PMA).

Cyclanthus bipartitus. V: Nibar. MED: Ground flower prepared as "Ina Kuamakalet" is drunk to avoid enlargement of the uterus in pregnant women. VS: Florpan No. 235 (PMA).

Dicranopygium crinitum. V: Aidu. MED: A whole plant infusion is used as a bath for weak children. This treatment is administered in a ceremony, where the old women smoke tobacco to enjoy the "Ina bundar." VS: Florpan No. 253 (PMA).

CYPERACEAE

Cyperus luzulae. V: Pina. MED: A whole plant infusion is used as a local bath for infections of the eyes and to avoid complications in childbirth. VS: Florpan No. 219 (PMA).

DILLENIACEAE

Davilla kunthii. V: Kuegi-dup. MED: An infusion of the branches and fruits is used as a bath for treating colic in children. VS: Florpan No. 282 (PMA).

Wedelia trilobata. V: Kanna. MED: A whole plant decoction is used as a local warm bath to treat skin diseases. VS: Florpan No. 234 (PMA).

EUPHORBIACEAE

Acalypha hispida. V: Mis Bonnu'gid. MED: A remedy prepared from ground flowers in the form of "Ina Kuamakalet" is drunk to facilitate childbirth. VS: Florpan No. 303 (PMA).

Croton fragrans. V: Ina'uirkina dup. V: The leaves and flowers prepared as "Inadibialet" is placed on the chest of the children to treat common cold. VS: Florpan No. 288 (PMA).

Croton lobatus. V: Ina'virkina dup purvignad. MED: A whole plant infusion is used as a local bath to treat skin diseases. VS: Florpan No. 290.

Chamaesyce hirta. V: Ua'ob Ina. MED: The leaves prepared as "Inadibialet" are used locally to treat acne. VS: Florpan No. 224 (PMA).

Jatropha sp. V: kala'sapi. MED: An infusion of ground seeds is drunk to induce vomiting. VS: Florpan No. 249 (PMA).

Manihot esulenta. V: Mam. MED: Cysts made by the insects on the leaves are prepared as "Inadibialet" and used to treat acne. VS: Florpan No. 300 (PMA).

GESNERIACEAE

Chrysothemis friedrichsthalana. V: Pirguak sipuguad. MED: A whole plant infusion is used as a bath to alleviate muscle aches and inflammation of the joints. It is also used for snakebites. VS: Florpan No. 265 (PMA).

Chrysothemis pulchella. V: Pirguak Kinnid. MED: A flower infusion prepared as "Ina Kuamakalet" is used from the sixth month of pregnancy to broaden the birth canal and to facilitate childbirth. VS: Florpan No. 301 (PMA).

GRAMINEAE

Coix lacryma-jobi. V: Oo'ina. MED: A root and fruit decoction prepared as "InaDibialet" is placed on the chest to alleviate a common cold. VS: Florpan No. 296 (PMA).

LABIATAE

Hyptis capitata. V: Ina etolo. MED: A remedy prepared from the young leaves as "Ina Kuamakalet" is drunk to treat snakebites. VS: Florpan No. 230 (PMA).

Hyptis Sp. V: Coke. MED: The whole plant is used as an excipient in the preparation of many dosage forms because of its aromatic odor. It is useful in "Kurgin Ina," which is a medicine for "Enlargement of the Mind," when the "brain is saturated." VA: Florpan No. 298 (PMA).

Ocimum canum. V: Bisep. MED: It is used as an excipient in some formulations for its aromatic odor. An infusion of the leaves and flowers is used locally as a bath for "Enlargement of Mind." VS: Florpan No. 245 (PMA).

Ocimum campechianum. V: Bisep Sapur. MED: It is used for its aromatic odor. VS: Florpan No. 293 (PMA).

LECYTHIDACEAE

Gustavia superba. V: Tupu. MED: A remedy prepared as "Ina Kuamakalet" is drunk for mental disorders. VS: Florpan No. 206 (PMA).

LEGUMINOSAE
Calliandra stipulacea. V: Udup. MED: A whole plant infusion is used as a bath for internal fever in weak children to whom the "nele" has diagnosed death. VS: Florpan No. 254 (PMA).

Prioria copaifera. V: Soila. MED: A stem bark infusion is used locally as a bath and a fruit infusion is drunk as a tonic and vigorant. VS: Florpan No. 209 (PMA).

LOGANIACEAE
Spigelia anthelmia. V: Ina Nasu. MED: A whole plant decoction is used as a local bath for skin infections and conjunctivitis. This plant is considered toxic. VS: Florpan No. 294 (PMA).

MALVACEAE
Hibiscus rosa-sinensis. V: Panab dutu, Panab dup. MED: An infusion of flowers is drunk during pregnancy for controlling the size of the uterus and as a bath for "mental saturation." VS: Florpan No. 243 (PMA).

Hibiscus schizopetalus. V: Pane'gid dutu. MED: A flower infusion is drunk to facilitate childbirth. VS: Florpan No. 232 (PMA).

Pavonia fructicosa. V: Mis Koo'gid. MED: A whole plant decoction prepared as "Ina Kuamakalet" is used as a bath to alleviate fever and common cold. VS: Florpan No. 231 (PMA).

Sida acuta var. *acuta*. V: Kuala. MED: A remedy from the decoction of the aerial parts prepared as "Ina Kuamakalet" is drunk as a tonic antipyretic and in cases of hair loss in children. VS: Florpan No. 246 (PMA).

MARANTACEAE
Calathea lutea. V: Urua'dili, Urua. MED: A flower infusion is used locally as a bath for weak children and to increase their learning capacity. VS: Florpan No. 287 (PMA).

MONIMIACEAE
Siparuna. V: Urgurgia dup. MED: An infusion of the leaves is used locally as a bath for fever in children. VS: Florpan No. 281 (PMA).

MORACEAE
Cecropia peltata. V: Niila. MED: A decoction of the inflorescence is used as a bath for headaches and "mental saturation." VS: Florpan No. 286 (PMA).

Dorstenia contrajerva. V: Ina niilaki'saila. MED: A whole plant infusion is drunk for snakebites and as a bath for muscle aches. VS: Florpan No. 267 (PMA).

MUSACEAE
Heliconia mariae. V: Tagar. MED: A fruit decoction is used as a bath for six days to improve blood circulation. VS: Florpan No. 211 (PMA).
Heliconia platystachys. V: Taggar Maku'gid. MED: A fruit infusion is used as a bath to facilitate childbirth. VS: Florpan No. 260 (PMA).
Heliconia vaginalis. V: Punur. MED: A whole plant decoction is drunk four times per day in cases of incorrect position of the fetus. VS: Florpan No. 223 (PMA).

PASSIFLORACEAE
Passiflora vitifolia. V: Sulegusep dup. MED: A fruit infusion is used as a bath for sixteen days for "mental saturation"; the patient should be admitted in a "surba"—a special room. VS: Florpan No. 216 (PMA).

PIPERACEAE
Piper auritum. V: Molinak'gid Bachar. MED: An infusion prepared from the inflorescence as "Ina Kuamakalet" is drunk to treat common colds. VS: Florpan No. 218 (PMA).
Piper hispidum. V: Bacher. MED: A leaf decoction is used to treat conjunctivitis, and a decoction from the inflorescence is used locally as a bath for muscle aches. VS: Florpan No. 213 (PMA).
Piper multiplinervium. V: Bachar dubaled. MED: An infusion of the young leaves prepared as "Ina Kuamakalet" is drunk for body and stomach aches. VS: Florpan No. 251 (PMA).
Piper tuberculatum. V: Ina tirki bachar, Inatisi bachar. MED: The ground inflorescence prepared as "Ina Kuamakalet" or a leaf decoction is drunk for liver pains. VS: Florpan No. 221 (PMA).

RHIZOPHORACEAE
Rhizophora mangle. V: Aili. MED: A ground root infusion is drunk by pregnant women to have a male child and as a bath for fractures of the bones. VS: Florpan No. 284 (PMA).

RUBIACEAE
Borreria laevis. V: Ina Pogiria. MED: A whole plant decoction is used as a warm bath for muscle aches and bone aches, inflammation, skin diseases, and acne. VS: Florpan No. 292 (PMA).

Coffea arabica. V: Ugurgia dup. MED: An infusion of the leaves is used as a bath for fever in children. VS: Florpan No. 289 (PMA).

Genipa americana. V: Sabdor. MED: A leaf infusion prepared as "Ina Kuamakalet" is drunk during two months beginning the third month of pregnancy to regulate the growth of the fetus. The fruit is rubbed on the skin to treat weakness in girls. VS: Florpan No. 217 (PMA).

Pentagonia pinnatifida. V: Kuaman dutu. MED: A root infusion is used as a bath for internal fever and the ground flower prepared as "Ina Kuamakalet" is drunk to facilitate childbirth and to induce menstruation. VS: Florpan No. 263 (PMA).

RUTACEAE
Citrus limon. V: Naras'chole. MED: The juice from the peels in single doses, every three days, is used by inhalation for common colds, breathing difficulties, and for coughs. VS: Florpan No. 233 (PMA).

SAPINDACEAE
Sapindus saponaria. V: Ina sasili. MED: An infusion of the ground fruits is used as a bath for skin diseases and colds. VS: Florpan No. 283 (PMA).

SELAGINELLACEAE
Selaginella sp. V: Naibe'uu. MED: A remedy from the leaves, prepared as "Ina Kuamakalet," is used for fever, weakness in children, muscle aches and eye infections. VS: Florpan No. 256 (PMA).

SOLANACEAE
Capsicum annuum. V: Kabur. MED: The ground fruit prepared as "Ina Kuamakalet" and the leaves as "Inadibialet" are used for serious ailments. VS: Florpan No. 291 (PMA).

Solanum lancaeifolium. V: Mogigia dup. MED: A whole plant decoction is used as a warm bath for muscle and stomach aches. VS: Florpan No. 226 (PMA).

Witheringia solanacea. V: Tinanquak'gid. MED: A whole plant infusion, the fruits and fresh leaves prepared as "Inadibialet" are used as a bath for body aches and skin diseases. VS: Florpan No. 268 (PMA).

STERCULIACEAE
Theobroma cacao. V: Ciamachi. MED: The stem bark prepared as "Ina Kuamakalet" is used to facilitate childbirth. This plant should not be given before sixth month of pregnancy because it may cause abortion. VS: Florpan No. 241 (PMA).

Herrania purpurea. V: Cuin'cia. MED: A remedy from the aerial parts, prepared in the form of "Ina Kuamakalet," is used to facilitate childbirth. It is said that "sticky" plants help facilitate childbirth. VS: Florpan No. 254 (PMA).

URTICACEAE
Urera laciniata. V: Take. MED: The urticant effect of the leaves is used to relieve body aches. VS: Florpan No. 279 (PMA).

VERBENACEAE
Lantana hispida. V: Uu'gua. MED: A flower infusion is used as a bath in cases of placenta retention. VS: Florpan No. 228 (PMA).

ZINGIBERACEAE
Dimeracostus strobilaceus. V: Pinnue barbat. MED: A flower and leaf infusion is used as a bath to improve blood circulation, while the inflorescence prepared in the form of "Ina Kuamakalet" is used to increase vitality in children. VS: Florpan No. 255 (PMA).

Acknowledgments

The authors wish to thank the Regional Program for Scientific and Technological Development of the Organization of American States and the Commission of the European Communities (Project CI★ 1.0505 ES [JR]) for supporting this research. We are also grateful to the Cuna Indians who let us share in their knowledge and to Emma Caballero for typing this manuscript.

Notes

1. Joly, L. G., Guerra, S. Séptimo, R., Solís. P. N., Correa, M., Gupta, M., Levy, S., and Sandberg, F. Ethnobotanical Inventory of Medicinal Plants Used by the Guaymi Indians in Western Panama. Part I. *Journal of Ethnopharmacology.* 20:145–71.

24

Searching for Plants in Peasant Marketplaces

GARY J. MARTIN
University of California–Berkeley

When we search for novel botanical resources in the tropics, how do we plan our itinerary? Primary forests and villages are tempting destinations, but perhaps a more cost-effective approach would be to begin in the marketplaces that are distributed throughout much of Latin America.

In these marketplaces, we find arrays of flowers, vegetables, fruits, dried herbs, and spices that are the first suggestion of the region's floristic richness and of its most useful plants. A great number of species sold are curative, an indication of the importance of herbal medicine in Latin America. Edible plants reflect the diverse origins of the local diet, which includes both native and introduced species, some domesticated and some collected from the wild. During ritual and religious seasons such as Christmastime, Easter week, and the Day of the Dead, ornamental plants are a showy element. A number of palm, grass, and woody species are sold as handcrafted utensils, including straw mats, baskets, brooms, and toys.

Tracing these diverse plants back to their zone of production, we find ourselves visiting all of the vegetation types of Latin America, from dry deciduous forests to humid tropical *selvas*. Among the buyers and sellers, there is a mix of several different social groups. Although one ethnic group may dominate market exchange, many other ethnic minorities are represented as occasional merchants or clients. Peasants are in abundance, urban dwellers are frequent visitors, and tourists are afoot.

Because of this rich cross-section of botanical, ecological, and social diversity, marketplaces appear to be a logical starting point in the search for novel botanical resources. But does this strategy provide results? A partial answer can be found in an ethnobotanical study that I have been carrying out with colleagues in Oaxaca, Mexico.

METHODOLOGY

When we began our study in 1985, there was no ethnobotanical inventory of the marketplaces of Oaxaca. Inspired by the observations of botanists and anthropologists who had previously visited the state, we set out to document the plants sold in the Central Valley and Sierra Norte markets. In the first stage of the project, I supervised three students from the Universidad Nacional Autonoma Benito Juarez de Oaxaca, who collected ornamental and medicinal plants in Oaxaca City markets from October 1985 to April 1986. Periodically I visited most of the principal and secondary markets of the Central Valley and Sierra Norte, making collections from April 1987 to March 1989; I was assisted in the summer of 1987 by a biology student from Baylor University. This market study was part of a larger project that includes systematic ethnobotanical inventories of selected Chinantec, Mixe, Mixtec, and Zapotec populations of the mountainous zones of Oaxaca (de Avila and Martin, 1990; Martin and de Avila, 1990).

In the markets, we concentrated on collecting plants, native or introduced, that are harvested in the state of Oaxaca. Our inventory thus excludes most of the plants that are produced only in other parts of Mexico and that are brought to Oaxaca from Mexico City's Mercado de Sonora, or from other external sources. We chose not to collect commonly cultivated food plants that showed no significant regional variation; for example, standard varieties of cauliflower and broccoli are grown in the Oaxaca valley, but do not form part of our collections. Apart from these exceptions, we feel that our survey includes a large majority of locally-produced native and exotic plants that are sold in the *plazas,* as marketplaces are locally called.

In all, we collected 651 voucher specimens, including 626 vascular plants (606 flowering plants, 8 conifers, 2 cycads, 10 ferns), 17 mushrooms, 7 bryophytes, and 1 lichen. Two hundred sixty specimens were live, whole plants that we pressed and dried, and have distributed to herbaria in Mexico and the United States. Three hundred ninety-one fragmentary species have been placed in jars or plastic bags; roots, crushed leaves, seeds, and other dry material were preserved as is, while fleshy material was pickled in 90 percent alcohol. One set of these vouchers has been distributed to the Jardin Botanico of the Universidad Nacional Autonoma de Mexico. Although the market collections are sometimes sterile and thus difficult to identify, we have succeeded in determining 98 percent to family (633 specimens), 92.5 percent to genus (602 specimens), and 67.5 percent to species (439 specimens).

Among these ethnobotanical collections, there are 300 species of vascular plants, which correspond to 256 genera and 97 families. Several families are well represented in the collection; there are 42 species of Asteraceae, 23 of Leguminosae, 19 of Lamiaceae, and 10 of Bromeliaceae. Nine additional families are represented by 5 or more species, while the remaining 84 families have between 1 and 4 species.

There is a wide range of uses for these plants; for the purpose of this study I have arbitrarily divided them into five general categories—medicine, food, ornament, condiment, and other uses. Although it is common for one plant to be used in multiple ways, I have classified the species according to their primary use. For example, I consider *Satureja spp.* as medicinal plants, but they are also used as a condiment and ornamental. According to this analysis, 32 percent of the plants are medicinal, 31 percent are used as food, 23 percent for ornamental purposes, and 8 percent as condiments. The remaining 6 percent are used for other, diverse purposes; some correspond to handcrafted household utensils such as brooms, mats, and baskets, while others are used for ritual purposes.

Although all of these plants are produced in the state of Oaxaca, 35 percent are exotics, which I define as plants that were introduced to the region since the arrival of the Spanish in Mexico. Among the 65 percent of plants that I consider native, there are several neotropical plants, such as peanuts and manioc, which were introduced in the prehispanic period. The proportion of exotic versus native plants varies over the different categories of use; we find the largest percentage of natives among the medicinal plants (76.1 percent), followed by food plants (59.6 percent), ornamentals (55.7 percent), and condiments (47.9 percent).

In the following section these categories are described in greater detail. Rather than listing the species in each category, I give examples of notable plants, commenting on their management and marketing. In the course of the discussion, I describe some of the techniques and issues that must be considered in marketing ethnobotanical studies.

FOOD PLANTS

Food plants illustrate particularly well the diverse ecological and ethnic zones represented in the market collections, and the local genetic diversity of domesticated plants. From the areas formerly covered by seasonal evergreen forest come bananas (*Musa spp.*) and citrus fruits (*Citrus spp.*), some of which are harvested by lowland Chinantecs from the Gulf Coast of Oaxaca. The cloud forest contributes *mamey (Pouteria sapota)* fruits, often brought by Sierra Mixe, and tepejilotes (*Chamaedorea tepejilote*), which I have seen being sold by Sierra Zapotec who inhabit humid zones. Sierra Zapotecs vend faba beans and other cold-weather crops, which they cultivate in dry lands in pine-oak forests over 2,000 meters above sea level. In the tropical deciduous forest, cultivated agaves are the source of *pulque,* a fermented drink that is consumed in many of Oaxaca's ethnic communities; I have seen it marketed by Zapotecs and Mixes in Sierra Norte markets. *Escontria chiotilla* and other cactus species from the semi-desert zone produce fruits that are carried to market by Zapotecs and mestizos from the central valleys and the Cabada.

MEDICINAL PLANTS

Marketing of medicinal plants has pushed species toward extinction in some areas of the world, most notably Africa, where many barks and roots are sold (Cunningham, 1990). We have detected no endangered medicinal plants in our collection, perhaps because the remedies mostly consist of leaves, flowers, and fruits of herbaceous species, which are less susceptible to depletion than species from which bark is stripped or roots harvested.

Although plant vendors can often provide information on which species are becoming rare, a better way to detect the sustainability of harvest is to visit the locality where the plants grow. Although not part of the market survey, I had the opportunity to interview plant gatherers and to see exploited stands of plants in the Sierra Norte. In some communities, a large number of people are dedicated to the production and selling of medicinals and condiments. San Juan Quiotepec, a Chinantec municipality of some 5,000 people, is famed for its herb merchants; approximately 50 to 60 families participate in this enterprise. All family members help in cultivating herbs such as chamomile (*Matricaria recutita*) or in gathering wild medicinals such as *te de tila (Ternstroemia sphaerocarpa)*. When a good stock is on hand and the herbs have been packed into small plastic bags, the adult males begin a trek that brings them to small communities in Oaxaca, Puebla, and Veracruz. The merchants go as far as Mexico City, where they purchase spices in the Mercado de Sonora, a clearing house for botanicals. These they sell on their way back to Quiotepec, where they begin the cycle again.

A complete inventory of their wares has yet to be taken, and we lack a study of the ecological impact of their harvesting activities. However, there are indications that some medicinals are produced in a sustainable way—chamomile is easily cultivated, and the *Bursera* trees that are tapped for copal resin appear to produce yearly with no ill effect. Other resources may be overexploited. The tila fruits, a popular remedy for calming nerves, are gathered in a dry tropical zone that forms the border between Quiotepec and Comaltepec, another Chinantec community. The municipal authorities of Comeltepec have jailed several Quiotepec Chinantecs, whom they accused of collecting tila fruits and other herbs on communal property. This may be occurring as the desired species become scarce within the municipal boundaries of Quiotepec.

ORNAMENTAL PLANTS

Exotic plants are sold for a variety of purposes in the markets, but they are nowhere more noticeable than in the flower stands. The native species—tuberose, *Milla biflora* lilies, and mock orange—make their presence known through scent, the heady perfumes giving away the location of florists. But the color display comes from exotics—South African calla lilies, European daisies, California poppies. These species teach a lesson in the introduction of new plants to the marketplace. In searching for unexploited species of commercial value, we eventually confront the problem of how to distribute the production. The presence of exotics gives us a clue that new plants have been introduced in the past, and the new arrivals become a part of the local useful flora.

Both native and exotic plants demonstrate the importance of recording seasonality in market studies. Dried medicinal herbs, condiments, and handcrafted utensils are available year-round, but food plants and ornamentals obey the laws of nature and culture. Wild orchids and magnolias have a defined flowering season, often coinciding with Easter, a prime time for selling ornamentals. Cultivated flowers may be produced all year, but there are times of high demand, such as Christmastime and All Saints' Day. When we visited the markets in these holidays, we concentrated on collecting the unique set of ornamentals that arrive from diverse parts of Oaxaca state.

CONDIMENTS

Among the condiments, we find a nearly equal mixture of native and exotic species. Although the spice merchants of the Oaxaca markets often place cinnamon and allspice side by side, the two species have quite distinct origins. Few people associate allspice *(Pimienta dioica)* with the mountains of Oaxaca, but it is a native species that ranges as well to the West Indies and Central America. Cinnamon *(Cinnamomum verum)* hails from Sri Lanka and Southwest India.

In the early 1980s, several governmental agencies combined efforts to promote the production of these species in the northern part of Oaxaca. Over the past several years, nearly 90,000 saplings of cinnamon were planted in the lowland Mixe region. Small plantations of allspice were promoted in the Zapotec region in the early 1980s. For the time being, the vendors of the Oaxacan markets can offer these domestically produced spices at a lower cost than imported sources, but the long-term scenario remains uncertain. The government has been providing a large subsidy since the inception of the project, and if this ends, the spice growers will have to rely on market prices as their only stimulus to continue. At present, the spices are sold to intermediaries who resell at a profit of at least 50 percent. Because the species are relatively new to the area, no agronomist is able to guarantee that the current production can be maintained over many years.

OTHER USES

There are many plant products that do not fall into the previous categories, most notably firewood, utensils, and ritual objects. Of these, firewood is the most ubiquitous. Some vendors come from impoverished villages where not much else is produced. Oakwood is preferred, and several species of *Quercus* are represented in our collections. Because it is preferred, oak is being overexploited; some mountain villages near the Oaxaca Valley will be completely deforested before the end of the decade. As oaks become increasingly rare, the demand for firewood will have to be satisfied by exploiting species from other vegetation zones. In the cloud forests, the local people consider a locally abundant tree, *Ticodendron incognitum*, to be as good as oak. This recently described species belongs to a new family of plant related to oaks and beeches; it is restricted to wet montane zones in Mexico and Central America (Hammel and Burger, 1991). In the absence of a sustainable agroforestry project in Oaxaca, this and other rare species could be threatened in the near future.

Most utensils are handcrafted from species that are managed in a sustainable way. Bowls are made from the fruits of *Lagenaria siceraria* and

Crescentia cujete, and baskets and mats are made from the vegetative parts of cattails (*Typha spp.*) and reeds (*Juncus spp.*).

However, when production turns from satisfying local needs to supplying the tourist trade, some species become quickly endangered. On any trip to the marketplaces, tourists are apt to be approached by the ubiquitous vendors of letter openers and other ornamental objects made from a yellowish-white, pliable wood. The ambulatory merchants are from Santa Cecilia Jalieza, a Zapotec village in the driest arm of the Oaxaca Valley; the wood is called *yagalán* and comes from *Wimmeria persicifolia,* a species of dry tropical forests. According to the villagers, the local populations of *yagalán* have disappeared, and the wood carvers are forced to buy the wood from more remote villages such as San Miguel Peras. When I interviewed a vendor from this community, she said that her sons were obliged to venture farther and farther from the village in search on *Wimmeria* trees.

Wimmeria and other species that yield wood for carving are increasingly rare as a result of the influx of tourists into Oaxaca. These species are sold locally, but any attempts at exporting them could increase demand and threaten wild populations.

The plants that received the least attention in the survey are those used for ritual: copal incense for the church, *Mucuna* seeds to ward off the evil eye, sprigs of basil to protect street vendors. Many of these plants are sold alongside medicinals in specialized stalls in the market. I believe that they are a less important element of market ethnobotany than is the case in Africa or even other parts of Latin America, but they deserve more attention in the future.

PLANTS OUTSIDE THE MARKETPLACE

Our ethnobotanical survey shows that a diverse group of species are being commercialized, and that marketplaces are a good place to begin to search for novel botanical resources. But where else can we look for useful plants, especially those not sold in markets?

For one, we can examine the archaeological record. Flannery (1986) and his colleagues have carried out extensive excavations in caves near the Central Valley, and they have described numerous plant remains. In this list, we find many plants sold in markets—the basic trio of corn, beans, and squash, as well other food plants such as *tejocote* (*Crataegis pubescens*) and nanches (*Malpighia spp.*). A number of species found in the cave sites are now considered as emergency foods; such plants as hackberry (*Celtis spp.*) and milkweed (*Asclepias spp.*) are only occasionally eaten in Valley villages and are never brought to market.

The archaeological data give a preliminary idea of how plant use has changed over the centuries. With a historical perspective we can discover which plants were formerly brought to market, but whose commercialization has been discontinued in recent years. A complete analysis of the change in marketing would require comparison to a previous market survey. Although no complete inventory exists for the Oaxaca marketplaces, we can refer to some observations that Richard Evans Schultes made during his fieldwork in northeastern Oaxaca, carried out in 1938 and 1939. Schultes (1941) notes 20 plants that he observed in Valley and Sierra Norte markets. Of these, we found 14 that are still being sold today; the remaining 6 have apparently disappeared from the markets. Schultes specifically mentioned the absence in the markets of two fruits, nanche (*Byrsonima crassifolia*), and zapote blanco (*Casimiroa edulis*). In our survey, we observed nanches in some Valley markets; the zapote blanco grows in the Valley but we have never seen it marketed.

Systematic ethnobotanical surveys in indigenous communities reveal other plants employed locally but not brought to market. Many of the plants that are used in Sierra communities can be found in Central Valley markets, but there also exists a large repertoire of botanical resources that are apparently never commercialized.

These other approaches indicate that collecting in markets is appropriate for identifying some but not all local useful plants. It is an important research tool, but not an exclusive approach; it must be complemented by archaeological and historical material, and its results must be compared with those of ethnobotanical research in neighboring communities.

DETECTION OF UNDEREXPLOITED PLANTS

Studies of market plants give one perspective from which to evaluate plants that may play a larger role in the development of peasant communities. By what criteria can we make a selection of which botanical resources are most likely to be successful candidates for increased production and marketing? I propose three dimensions—economy, ecology, and culture.

The most obvious considerations are economic: Is the market for the plant expandable, and is it relatively permanent? Does the market provide a suitable replacement value for the human labor and resource depletion that increased exploitation implies?

The promoters of many projects begin to assess economic viability well after production has begun. For example, a loose cooperative of Chinantec speakers increased production of vanilla orchids as a way to finance various cultural products in the lowland Chinantla. The plants flourished, but the cultivators are now looking for ways to compete with other producers on the international market. A similar quandary exists for the spice producers, mentioned above, who currently sell their harvest to intermediaries.

Optimists may counter that, given the current interest in forest conservation and cultural survival, some part of the production can be sold at advantageous prices through creative marketing. In Europe and in the United States, there is a demand for the purchase of products that putatively contribute to solving the plight of disappearing forests and indigenous cultures. Since many forest products can also be sold under an organic label, their value is increased.

The lesson to be learned is that markets based on political ideology are ephemeral, and that competing in more established markets thrusts us into the world of commodity exchange, where people may be indifferent to ethnic and ecological considerations. Many nontimber forest products may catch our attention as potential resources, but few will be competitive in international markets.

Beyond economy, we must look at ecological restraints. Since the publication of the paper by Peters et al. (1989) on the evaluation of botanical resources of an Amazonian rain forest, several biologists have expressed skepticism about the sustainability of the production of many would-be crops (Bodmer et al., 1990). Would increased collection of fruits and seeds interrupt the life cycles of certain species, leading to diminished populations in the future? Would extensive gathering of bark, roots, and foliage kill off a large number of individual plants, pushing some species toward extinction?

Overexploitation is damaging not only to the local economy and ecology, but also to the culture and folk knowledge of the local people. Culture is a new word in the vocabulary of many ecologists and economists who are accustomed to thinking about profitability and diminishing returns in a linear fashion. Anthropologists have added a curve to the equation—what is profitable and sustainable may not be desirable from a cultural point of view.

When thinking of plants that may have an international market potential, *rosita de cacao,* the fragrant flower of *Quararibea funebris,* comes to mind. My colleagues and I have thought that the 'Cocoa rose' could become a popular condiment, because we had often enjoyed its maply flavor in tejate, a local drink made from corn, mamey seeds, and cocoa. Its production is sustainable; the tree, which is from the lowland tropics, has been cultivated in some villages of the Central Valley since prehispanic times. Our concern is for the lifestyle of Oaxaca. Tejate is a fixture in valley markets, finding its place alongside the *aguas frescas* made from locally produced fruits.

Schultes found 'rosita de cacao' in the markets of Oaxaca fifty years before we encountered it. If we were to promote its commercialization in the United States, would it and tejate still be found in marketplaces fifty years from now? In Huayapam—where most of the valley *rosita* is produced, and tejate is a characteristic part of the cuisine—villagers already complain about the rising cost of ingredients. *Mamey* and cocoa seeds, brought from other ecological zones, are increasingly expensive. If the *rosita de cacao* were to become a coveted spice, then tejate consumption in Oaxaca might become a thing of the past.

Although the loss of one traditional drink would not worry even a nutritionist, the loss of an entire cuisine would have disastrous consequences. In this sense, *Quararibea funebris* is not alone—local plants and diet are inextricably linked, and in many cases greater affluence has led to poorer nutrition for indigenous communities.

THE ETHICS OF MAINTAINING REGIONAL TRADE AND COMMUNICATION

It is fashionable to speak of these three dimensions—economy, ecology, and culture—under the broad umbrella of ethics in ethnobiology. As concerned academics, we must consider the long-term effects of promoting the commercialization of botanical resources. By looking for external buyers for local products, we encourage peasants to become further entrenched in the world economic order. For the Mexican peasants, this is not a novel enterprise but a traditional element of their local economy that grows and ebbs with the health and caprices of the international economy. As part of their household economy, they have been providing goods to national markets since the prehispanic period; the arrival of Europeans caused them to participate on a world-wide scale. Yet there is another side to peasant production—the goods that are for domestic use, or that are exchanged in regional markets. In contrast to the produce bound for national and international markets, these goods play an important role in maintaining the well-being of the local people.

Should we encourage local self-sufficiency or production for export?

It is curious that just as ethnobotanists are proposing the commercialization and export of botanical resources, some other colleagues, refugees from biotechnology, are moving in the opposite direction. Martha Crouch, who was conducting research on plant embryos and pollen development, commented recently, "In April 1990 I wrote an essay describing the destructive effects of my own research—helping to produce more uniform oil plants which would ultimately undercut Third World economies—which stimulated many scientists to write me about their own similar concerns. Today I am working to change agriculture back towards decentralized, non-export systems of local food production. Until food is removed from the realm of global commodities, new knowledge about plants will always be used against nature and against the poor" (Crouch, 1991).

After a few years of wrangling with commodity exchange, I wonder if ethnobotanists will be expressing the same opinion. Even as we look for exportable products among our ethnobotanical inventories, we should promote the regional exchange and consumption of plants. For each superstar species that actually brings a cash income to peasants, we will find dozens of plants that feed, cure, and shelter the local population. Although most of these species will never find their way out of villages and into the international marketplace, their impact on the standard of living goes well beyond any monetary value they might yield.

This gets us out of the world of economics and into the realm of cultural promotion. We need to support initiatives—newsletters, radio shows, traveling exhibits, ethnobotanical gardens—that encourage the expanded use of plants among rural and urban populations. This would reinforce the traditional lifestyles, and spread the benefits of local knowledge in a noncommercial way. Marketplaces are an appropriate site for this promotion, providing a diverse and curious public eager to learn about new foods, medicines, and other products. They are not only fertile territory for discovering traditional uses, but also the proving grounds for the introduction of new plants.

Ethnobotanists have an important role to play in this process—not as commodity brokers, but as researchers. Before we venture into the regional *promotion* of a species, we must scientifically validate its properties by carrying out phytochemical and other analyses. We must survey the different social and ethnic groups in the area to see how widely the plant is currently used, and we must study the potential of producing the plant in a sustainable way. After taking these steps, we can participate with local people in promoting regional exchange of plant products, and in discovering which species have the potential of being commercialized on a wider scale.

References

Bodmer, R., T. Lang, and L. Moya. 1990. Fruits of the Forest. *Nature* 343:109.

Boom, B. 1990, Giving Native People a Share of the Profits. *Garden* 14:26–31.

Bye, R., Jr., and E. Linares. 1983. The Role of Plants Found in the Mexican Markets and Their Importance in Ethnobotanical Studies. *Journal of Ethnobiology* 3:1–13.

Crouch, M. 1991. Confessions of a Botanist. *The New Internationalist* 217:21.

Cunningham, A. B. 1990. *African Medicinal Plants: Setting Priorities at the Interface between Conservation and Primary Health Care*. Report for WWF project 3331.

de Avila, A., and G. J. Martin. 1990. Estudios Etnobotanicos en Oaxaca. In E. Leff (ed.), *In Recursos Naturales, Tecnica y Cultura*. Estudios y Experiencias para un Desarrollo Alternative. Mexico City: CIIH-UNAM.

Flannery, Kent U. (ed.). 1986. *Guila Naquitz, Archaic Foraging and Early Agriculture in Oaxaca, Mexico*. New York: Academic Press.

Hammel, B., and W. G. Burger. 1991. Neither Oak nor Alder but Nearly: the History of Ticodendracea. *Annals of the Missouri Botanical Garden* 78:89–95.

Martin, G. J., and A. de Avila. 1990. *Exploring the Cloud Forests of Oaxaca*. WWF Reports, October/November/December issue.

Messer, E. 1978. Zapotec Plant Knowledge: *Classification, Uses and Communication about Plants in Mitla, Oaxaca, Mexico*. Ann Arbor, Michigan: Memoirs of the Museum of Anthropology, no. 18. University of Michigan.

Peters, C. M. 1990. Plenty of Fruit but No Free Lunch. *Garden* 14:9–13.

Peters, C. M., A. H. Gentry, and R. C. Mendelsohn. 1989. Valuation of an Amazonian Rainforest. *Nature* 339:655–56.

Posey, D. 1990. Intellectual Property Rights: What Is the Position of Ethnobiology. *Journal of Ethnobiology* 10:93–98.

Schultes, R. E. 1941. *Economic Aspects of the Flora of Northeastern Oaxaca, Mexico*. Unpublished Ph.D. Thesis, Harvard University.

25

The National Cancer Institute's Plant Collections Program: Update and Implications for Tropical Forests

Douglas C. Daly
The New York Botanical Garden

On September 1, 1986, the Natural Products Branch of the National Cancer Institute (NCI) began a new program to survey the biological diversity of the tropical world for new chemotherapeutic agents for the treatment of AIDS and various cancers. Five groups of organisms are being sampled: plants, marine organisms, cyanobacteria, fungi, and protozoa. The New York Botanical Garden (NYBG) is one of three institutions completing the fifth year of its work in the tropics for the Plant Collections Program, which is screening extracts of thousands of species of higher plants from the tropics for anti-HIV and anti-cancer activity. Each institution is responsible for collecting at least 1,500 bulk samples between 0.5–1 kilograms dry weight of the bark, roots, twigs, and other organs of as many species of plants as possible. Each sample corresponds to a herbarium voucher collection that confirms the identity of the species. NYBG is operating the program in the New World tropics, the University of Illinois/Chicago in Asia, and the Missouri Botanical Garden in Africa.

This effort replaces an earlier program that functioned between 1960–1980. The new screening program differs significantly from its predecessor in three important respects. First, new technology at NCI has produced screens with greater sensitivity and specificity: the more generalized murine (mouse) systems utilized from 1960–1980 have been replaced by a series of specific human cancer-cell lines, as well as AIDS-infected cell cultures. A number of human cancer-cell lines are still being developed. Second, the material that is not extracted immediately and expended in the assays is being archived at $-20°$ Celsius. Therefore, as cancer research progresses, future assays will have recourse to an invaluable sampling of the chemical diversity of the tropics without duplicating the costs in what would be a very difficult and expensive effort to retrieve all the same material. Third, rather than commission the vouchered samples from a constellation of sources, NCI issued contracts to a limited number of institutions that have wide-ranging research programs on a given continent, thus centralizing the information-gathering effort.

RESULTS

Table 25-1 summarizes the production and preliminary results of the NCI screens as of May 1, 1991. Because of the pressing nature of the AIDS epidemic, priority has been given to that screen. Over 800 extracts have shown some anti-HIV activity, incluing 436 plant extracts. However, many of these are aqueous extracts that have been shown to contain compounds that manifest strong *in vitro* anti-HIV activity but are of no interest as clinical candidates. There are probably 50 to 100 extracts in various stages of fractionation, and some novel agents have been isolated.

The cancer screens began operating more recently. Approximately 60 extracts have shown interesting activity, and about half of these are from plants. Some 30 of the 60 extracts are being studied at present.

In summary, a number of extracts have shown anti-HIV or anticancer activity, and between 80 and 130 extracts are being studied in

TABLE 25-1   *Preliminary Results of the National Cancer Institute Screen*[1]

	Total	Plants
No. of samples received	30,415	23,000
No. of extracts prepared	79,375[2]	26,504
No. of extracts put through the AIDS screen	28 700	13,032
No. of extracts put through various cancer screens	5,876	800
No. of extracts being examined further (e.g., fractionation)	80–130 (including plants)	—

[1]Information supplied by G. Cragg, May 1, 1991.
[2]In general, two extracts (one organic and one aqueous) are prepared of each organism. For fungi, two extracts are prepared per culture.

more detail. It must be emphasized that these preliminary *in vitro* results are only the first step in a long and complex process of drug development that includes screening, rescreening, evaluation of specificity (versus general cytotoxicity), fractionation and chemical analysis, investigation of other properties, isolation and/or synthesis, testing for *in vivo* activity, toxicity studies, pharmacokinetic analyses, and clinical trials. NCI estimates that of every 10,000 extracts screened, fewer than ten will reach clinical trials.

IMPLICATIONS FOR TROPICAL FORESTS

Based on the above numbers, it is likely that novel therapeutic agents will be discovered in the tropics as a result of the Plant Screening Program, and it is important to anticipate the possible consequences of such a discovery. At the far end of the drug development process, a recent success story from the temperate zone gives cause for hope as well as concern. Taxol is a compound that has shown remarkable activity against ovarian cancer (Einzig et al., 1990), and preliminary results indicate great potential in the treatment of breast cancer (Christian, 1990). However, the taxol used to date has been obtained from the very low concentrations that occur in the bark of *Taxus brevifolia,* a slow-growing tree native to the northwestern United States. Just the amount of taxol needed to complete the clinical trials could have an unacceptable impact on existing populations of *T. brevifolia.* Moreover, its distribution apparently coincides roughly with that of an endangered species of bird, the spotted owl. Consequently, the NCI is now supporting research to provide alternative sources of taxol, including tissue culture, synthesis, semi-synthesis from precursors and analogs, and large-scale cultivation of *T. brevifolia* and related taxa.

This success story-in-progress has important implications for what should be expected—or desired—from the current NCI program in the tropics. At the outset, it must be stated that production of a new chemotherapeutic agent for treatment of cancer or AIDS may not be compatible with extractive economies, because sustainable harvest from wild populations of most species may not be able to satisfy the demand for the raw material. The diseases currently being examined in the screening program are among the more common and recalcitrant cancers in the developed world, and AIDS is epidemic in many areas of both the developed and developing worlds. Any successful treatment will generate enormous demand. The case of taxol provides an example of how this demand can sometimes represent a threat rather than a boon to the habitats that give rise to new medicines. Governments and pharmaceutical companies cannot always be expected to behave in an ecologically enlightened manner.

The discovery of a successful new chemotherapeutic agent presents an array of possible benefits for and dangers to tropical forests. The consequences depend on several factors involving the ecology of the species and the nature of the chemical compound: 1) whether the compound can be synthesized (or semi-synthesized from precursors or analogs), 2) whether the compound occurs in low or high concentrations in the plant, 3) how much of the compound is required for a course of treatment, 4) whether the species occurs in primary or secondary vegetation, 5) whether the species occurs in dense or sparse populations, and 6) whether the species can be brought into cultivation.

If the compound can be synthesized in the laboratory at reasonable cost, harvest from cultivated material or wild populations will be irrelevant unless there are local uses for the species. When synthesis is prohibitively costly or difficult, harvest or cultivation may be chosen if production from the original plant source is adequate, consistent, and relatively cost-effective. This is more likely if the compound occurs in high concentrations in the plant and/or if only a small amount of the compound is required for a course of treatment. To meet any significant demand, however, production from plant material will be feasible only if the species occurs naturally in dense populations, or if it can be brought into cultivation without much difficulty.

Domestication will certainly be attempted for any species that is the source of a successful chemotherapeutic agent. In fact, species that lend themselves to cultivation tend to be taxa that occur naturally in dense populations. These are usually species of secondary vegetation or gaps, although a number of primary tropical forests are oligarchic, that is, they are dominated by few species that occur in high densities. Production via sustainable harvest from wild sources may be considered if the following conditions are met: the compound cannot be synthesized, the species cannot be cultivated or brought into tissue culture, the compound occurs in very high concentrations in the plant, and the species occurs naturally in high densities. If not, the plant populations and possibly the habitat will be in danger of being decimated by destructive harvest practices.

Large-scale cultivation may not exert pressure on the forest if plantations are located in degraded pastures, although this is extremely difficult to guarantee. Cultivation may be compatible with primarily extractive economies if carried out on a small scale. It should be emphasized that any production system that benefits extractive economies will help to maintain occupation and utilization of the forest by people who are more inclined to treat it responsibly.

Any new chemotherapeutic agent that results from the current NCI program can serve as a powerful argument for conservation. When one ceases the hand-waving about the potential value of tropical forests as sources of new medicines and novel substances of real value are found, the argument becomes more compelling. It becomes still more compelling if the source is a rare tree in primary tropical forest.

The same arguments can be used to defend forest cultures, which are far more endangered than the forests. NCI's first screening program found that plants indicated by local peoples as being medicinal, toxic, or anthelminthic were between two and five times more likely to score a "hit" in the assay (Spjut and Perdue, 1976), and preliminary data from the current program indicate similar results (Balick, 1990). It should be noted that ethnobotanical research can help to conserve plant lore within the culture by showing the younger, more acculturated generations that the knowledge of their elders is highly regarded in the outside world.

Unfortunately, the forests as well as the local peoples and the countries hosting this research need more than the promise of economic returns that in most instances may materialize only ten years in the future, when not only the plants but the peoples who manage them and first learned their uses may already have disappeared. For this reason, NYBG and several other research institutions began working with NCI to structure the program so as to reflect what we hope is a more enlightened approach to medicinal plant research. Environmental organizations and pharmaceutical companies have also become involved in these discussions. In general, research programs that seek cures for developed-world diseases should include some forms of up-front benefits to local peoples and institutions, ethical legal agreements concerning royalties and drug development, technology transfer to and training opportunities for scientists in the countries of origin, and when possible assistance with combatting tropical diseases. We have been able to make contributions in all but the last area, which is not within NCI's purvey. Much more progress is needed in conceiving and developing mechanisms to provide up-front benefits and royalties to local peoples, especially to nonfederated tribal groups.

OTHER RESULTS

In addition to its more straightforward value in medical research, the NCI Plant Collections Program has already produced other kinds of results. It has supported important botanical exploration, with the attendant value for traditional conservation strategies. It has supported botanical research programs of our colleagues in tropical countries, including training and institution-building. It has encouraged and supported development of valuable ethnobotanical field projects. This NCI program has operated in some 24 tropical countries, including 13 in the Neotropics. It has generated nearly 20,000 botanical collections and a total of probably about 80,000 herbarium specimens. These collections have included more than 40 species new to science, including a new genus from Africa and more than 30 new species in the Neotropics. Organs of many of these new taxa have been entered in the screens.

CONCLUSIONS

The current NCI Plant Collections Program is now in the fifth and final year of its first phase, and it is expected to continue for at least another five years. Given the the time-scale of drug development and the fact that more screens are being developed, the program is still in an early stage. A number of positive preliminary results have been obtained and are being pursued.

The discovery of a new chemotherapeutic agent through this program will have positive as well as potentially dangerous consequences for tropical forests. The discovery in itself will constitute a powerful argument for conservation. Unless the compound can be synthesized, however, most means of production are likely to put additional pressure on the forest through either destructive harvesting or large-scale cultivation. The implication is that synthesis is probably desirable in many instances, and that at least for research on AIDS and common cancers, the tropical forest should be treated as a source for new medicines, not as a place for production of these medicines.

Discovery of new anti-AIDS and anti-cancer compounds from tropical forest plants can benefit local peoples and scientific institutions in the countries of origin through contracts and licensing agreements that would provide royalties to them some ten years hence, but the needs of the forests, of forest peoples, and of scientists in tropical countries are much more immediate. For that reason, the NCI program has been structured to provide training opportunities for botanists and chemists in the countries of origin, to assist in the development of the host institutions, and to support the field programs of botanists at these institutions. Without a doubt, one of the most positive results of the NCI program is the debate it has fostered and the progress it has stimulated in the ethics of medicinal plant research.

Acknowledgments

I am grateful to Gordon Cragg of the National Cancer Institute's Natural Products Branch for the information he generously provided. James Miller of the Missouri Botanical Garden and Djaja Soejarto of the University of Illinois at Chicago also shared information about their institutions' programs.

References

Balick, M. J. 1990. Ethnobotany and the Identification of Therapeutic Agents from the Rainforest. In *Bioactive Compounds from Plants*. Ciba Foundation Symposium 154.

Christian, M. C. 1990. *Taxol Phase II and III Trials and Future Plans*. Talk presented at "Workshop on Taxol and *Taxus:* Current and Future Perspectives," sponsored by National Cancer Institute at Bethesda, June 26, 1990.

Einzig, A. I., P. H. Wiernik, J. Sasloff, S. Garl, C. Runowicz, K. O'Hanlan, and G. Goldberg. 1990. *Phase II Study of Taxol in Patients with Advanced Ovarian Cancer*. Talk presented at "Workshop on Taxol and *Taxus:* Current and Future Perspectives," sponsored by National Cancer Institute at Bethesda, June 26, 1990.

Spjut, R. W., and R. E. Perdue, Jr. 1976. Plant Folklore: A Tool for Predicting Sources of Antitumor Activity? *Cancer Treat. Rep.* 60(8): 979–85.

26

Pharmaceutical Discovery, Ethnobotany, Tropical Forests, and Reciprocity: Integrating Indigenous Knowledge, Conservation, and Sustainable Development

STEVEN R. KING
Shaman Pharmaceuticals, Inc.

Today, 500 years after the first encounter between the Old World and the New World, the global pharmacopeia still traces many of its most widely utilized medicines to the practitioners of ethnomedicine. Advances in the medical sciences, especially pharmacology, have been very closely connected to the knowledge of native people about the therapeutic value of plants.

Plants formed the basis of many healing arts in Europe at the time of the first journey by Columbus in 1492. The King of Spain retained botanists and pharmacists—often one person performed both duties—to identify, collect, formulate, and administer plant medicines for the royal family. The search for new nontimber forest products to add to this pharmacy was one of the many motivations for commissioning the journey of Columbus to discover a new route to the Indies.

At the same time, indigenous people of the New World had developed complex and sophisticated knowledge systems about the use of plants for medicine, food, and hunting. In Mesoamerica, Nahuatl-speaking Aztec healers classified and utilized a wide array of plants for medical purposes. Much of this knowledge was depicted in prehispanic painted books known as codices. Nearly all of these prehispanic codices were destroyed by Cortez and the Holy Inquisition. Ironically, the Spanish government and Church commissioned scholars to document as much detail as possible about the pre-Columbian life of the Mesoamericans. The most notable of these works are the Florentine Codex, by the Franciscan Friar Benardino de Sahagun, the Badianus Manuscript, created by two indigenous people, and the Codex Mendoza, which contains lists of plants and other items paid as tribute to the Aztec empire (Williams, 1990).

From the northern regions of the continent to Tierra del Fuego, indigenous people were utilizing a vast diversity of plants for medical purposes. In the Andes the Inca performed trypenations, head surgery—with a very high recovery rate—for a number of health problems (Verano et al., 1991). The native people of Amazonia discovered and combined a wide array of plants as medicine, including *Chondrodendron tomentosum,* one of the many plants utilized in arrow poisons known as curare. Today this species is widely used in surgical procedures as the source of the skeletal muscle relaxant tubocurarine.

One of the first gifts of knowledge from native people was "Indian fever bark," obtained from the bark of the several species of *Cinchona* trees. This well-known source of the antimalarial drug quinine was later renamed "Jesuit fever bark." The popularity of this bark from South America was extensive in Europe. Thousands of kilos of bark were shipped to Spain and thousands more awaited transport in Pacific coastal ports. The intense demand led to the commissioning of several scientific expeditions to examine the variation in sources and potential for increased supply. One of these expeditions included the botanists and pharmacists Ruiz and Pavon, whose work was part of the foundation plant exploration in Latin America. Nevertheless, this intense demand for quinine created one of the first plant conservation problems in tropical America. If this species had not been transferred to another part of the world for cultivation, it might have been extinguished in its natural range.

There are many other examples of medicines derived from indigenous knowledge about plants, including the amoebocide and emetic drug emetine, obtained from the roots of *Cephaelis ipecacuana.* One of the world's most important local anaesthetics, cocaine, derived from the leaves of *Erthroxylum coca,* was utilized historically and is still used today as medicine by thousands of people in the Andean region of South America. Pilocarpine, a drug used to treat glaucoma, is derived from the plant *Pilocarpus jaborandi* and was utilized medicinally by indigenous people in Brazil.

This impressive record of discovery is in no way complete. Less then 2 percent of the 90,000 plant species in the Neotropics has been examined for pharmacological activity. The tremendous biodiversity of tropical forests has led to the creation and strategy of Shaman Pharmaceuticals, Inc., a venture-capital supported traditional pharmaceutical company focusing on the creation of ethical pharmaceutical products.

The goal of Shaman Pharmaceuticals is to discover and develop novel pharmaceuticals from higher plants, especially antifungal, antiviral, and sedative/analgesic plants. We are working with ethnobotanical knowledge about the enormous biodiversity of plant life that exists in the tropical forests of the world. As part of our strategy, we are combining the disciplines of ethnobotany, isolation chemistry, and pharmacology to create a more efficient drug discovery process, then we are using mass screening and genetic engineering. We anticipate that the unique structural chemistry of these plants will lead to the discovery of new prototype compounds that act in the body on previously undiscovered receptors and body pathways (Figure 26-1).

This is significant because the bulk of new drugs approved each year is not novel, nor is it made up of prototypical chemical entities. In such instances, these drugs could provide modes of treatment currently unknown in the practice of medicine.

There is ample evidence that this approach will yield new candidates for the development of therapeutic agents with new mechanisms of action. Several novel chemical entities are being evaluated in human volunteers for antiviral activity, and the sources of these compounds are

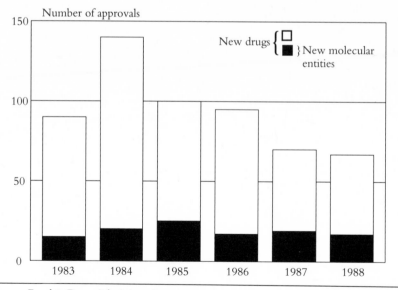

Source: Food & Drug Administration

FIGURE 26-1 Only a small fraction of drugs approved are totally new compounds.

plants utilized in Asian and African traditional medicine (Vlietinck and Vanden Berghe, 1991). In order to increase the interaction between scientific disciplines we are employing new tools and methodologies from the laboratory to the field research phase.

We have been utilizing a more direct interaction between ethnobotany and Western medicine. During field research conducted in January 1991, Dr. Charles Limbach and I utilized a color atlas of clinical dermatology (Fitzpatrick et al., 1983) with photographs as a cross-cultural tool to discuss plants utilized to treat fungal diseases. Using these images we were able to carry out discussions with many different people independently. We utilized images of common fungal skin infections that were readily recognizable, including athlete's foot (*Tinea pedis*) and fungal infections of the hand, head, and body (*Tinea manus, Tinea capitis,* and *Tinea corporis*). This method had been pioneered by Dr. Limbach while working in the Ecuadorian Amazon in 1989, and yielded excellent results in our fieldwork as well.

Through this combination of new and standard methods we collected more detailed data of greater use in determining the most appropriate chemical processing and pharmacological screens to use for the evaluation of the plants. There have been a number of scientists that advocate this combination of a medical person and ethnobotanist working together in the field (Balick, 1990; E. Elizabetsky, pers. comm.).

This type of detailed interplay in the field, combined with advanced chemistry and pharmacology, has yielded a high proportion of active compounds in our focused antiviral, antifungal, and sedative/analgesic screens. We have found that 50 percent of our samples have shown activity. We have also found that 74 percent of our active samples correlated with the original ethnobotanical use (Figure 26-2). At this time we are working on several of these within our therapeutic focus.

Discovering, developing, and marketing new therapeutic compounds are only part of Shaman's goals. We are also interested in creating opportunities for the development of sustainable natural product supply industries. Most of the compounds that we are investigating have complex chemical structures, which makes them difficult or impossible to synthesize on a commercial scale. A North American example of a compound whose chemical structure has defied commercial synthesis is that of Taxol, derived from the Pacific Yew, *Taxus brevifolia*.

As a company, Shaman Pharmaceuticals is committed to working out sustainable harvesting methods for the plant products that we are investigating. Preliminary research on sustainable harvesting methods for a specific extractive forest product is now being undertaken in three countries. In many cases nondestructive or renewable extractive methods

Number of species Collected	Number Awaiting Screening		Number Screened for Some Type of Activity	
104	46 (44%)		58 (56%)	
Of the 58 Species Screened:	Number of Samples	% of Total Collections	% of Total Screened	
Samples lacking activity in screens applied	23	22%	39%	
Samples having activity in 1 or more screens	35	34%	60%	
Of the 35 Active Samples	% Correlated with Folk Use		% Not Correlated with Folk Use	
	74%		26	

FIGURE 26-2 SUMMARY OF SCREENING RESULTS.

must be developed for a species that has previously been destroyed as part of regular harvesting methods. Sustainable harvesting methodology is critical for the local economy, to conserve the integrity of the ecosystem, and to conserve the genetic diversity of the species being commercialized.

As part of this strategy we are especially interested in plant sources that are common in secondary forest habitats, many of which are already utilized as medicine. They are often classic pioneer species. Supply industries based on plants with these qualities can reduce pressure on nearby primary forests and provide income.

The above-described strategy and goals of developing therapeutic agents from plants are part of an overall picture that is still incomplete. In the past, reciprocal benefits for indigenous and local people have not been provided, even though they are primary sources of the invention. Shaman Pharmaceuticals has created structures that are designed to address the ethical, financial, and conservation issues associated with indigenous knowledge and tropical plant genetic resources.

We have created a nonprofit conservation organization called The Healing Forest Conservancy. This organization, founded out of our firm belief that important new medicines can be discovered from plant materials, is dedicated to providing a structure in which indigenous people can participate and share responsibility for sustainable development and/or management of natural resources that are part of their cultural heritage. It is also dedicated to maintaining global plant biodiversity, especially those plants that have been used traditionally for medical purposes.

Through this organization a portion of the profits generated by commercialization of plant-derived compounds will be distributed to those appropriate organizations in countries that participate in the process of plant collection and other collaborative activities. The other objectives of this organization include: 1) working with local tropical forest conservation organizations to develop conservation programs to manage biological resources more effectively, 2) working with indigenous support organizations to facilitate the sustainable production and marketing of rain forest products, 3) promoting the sharing of knowledge among indigenous people, 4) producing multilingual publications for use among indigenous communities. The organization will also collaborate with other conservation organizations to provide accurate scientific information on endangered natural habitats, species, or populations of plants that are or may be important local sources of medicine.

A portion of the goals of the Healing Forest Conservancy will be realized over time as Shaman Pharmaceutical, Inc., initiates the process of product development. We are also committed to providing direct and immediate reciprocal benefits to indigenous people and the countries in which they live. We have asked the people with whom we have worked what they need and want. They told us very clearly what they need. One of the most frequent responses was access to health care.

In response, we provided health care to 60 individuals in Amazonian Ecuador. Dr. Charles Limbach, partially supported by the Healing Forest Conservancy, provided medical care to a wide array of people in two Waorani and one Shuar Indian community. Among the Shuar, Dr. Limbach treated 30 children that were experiencing a widespread epidemic of whooping cough.

The people of these two communities also expressed their desire for better access to health care facilities, especially in cases of snakebite and unfamiliar diseases. Individuals also expressed a desire to have access to forms of work that would provide them with local currency. These individuals pointed out that hospitals, merchants, and other services and products that they needed all required currency for payment. Addressing these needs should be part of any project that is working to develop new pharmaceutical products based on indigenous knowledge.

CONCLUSION

The challenges facing the sustainable harvest and marketing of rain forest products are numerous. There are no easy solutions to the many problems that can arise as old or new supply industries are developed. It seems quite clear, however, that people in tropical forests need viable economic alternatives to clearing and burning primary forests for short-term nonsustainable agricultural practices. It is also evident that the people of tropical forest regions are searching for feasible options to enable them to meet their basic needs of food, clothing, shelter, and health care.

There is no doubt that many options and products will be necessary to provide viable options to the people in these regions. No single sustainable pharmaceutical supply industry or 20 new foods or cosmetic products or whatever will be able offer enough alternatives to deforestation. We are working with a highly diverse cultural, physical, and political landscape. Countries with tropical forest habitats will need a diversified portfolio of potential products on several scales to meet the needs of people, the ultimate stewards of the tropical forests of the world.

The development of pharmaceutical products can be a vigorous part of a global conservation strategy while simultaneously acknowledging the ethical obligation of reciprocal benefit for indigenous people. Native people have been one of the most valuable sources of knowledge for the process of drug development. These people and their knowledge about the chemistry of tropical forests have been and will continue to be invaluable to our global pharmacopeia.

It is time to change the trends of the past five centuries, to implement a process that will provide benefits for local people that will enable them to manage wisely the genetic resources of their tropical forest with the *next* five hundred years in mind.

References

Balick, M. J. 1990. Ethnobotany and the Identification of New Drugs. In D. Chadwick and J. Marsh (eds.), *Bioactive Compounds from Plants*. Ciba Foundation Syposium 154. Wiley & Sons.
Fitzpatrick, T. B., M. K. Polano, D. Suurmond, and R. A. Johnson. 1983. *Color Atlas and Synopsis of Clinical Dermatology*. New York: McGraw Hill, Health Professions Divisions.

Verano, J., and D. H. Ubelaker. 1991. Health and Disease in the Pre-Columbian World. In H. Viola and C. Margolis (eds.), *Seeds of Change.* Washington, D.C.: Smithsonian Institution Press.

Vlietnick, A. J., and D. A. Vanden Berghe. 1991. Can Ethnopharmacology Contribute to the Development of Antiviral Drugs? *Journal of Ethnopharmacology,* 32:141–53.

Williams, D. 1990. A Review of the Sources for the Study of Nahuatl Plant Classification. *Adv. Econ. Bot.* 8:250–70.

V

Reaching International Markets

27

The Brazil Nut Industry—Past, Present, and Future

Scott A. Mori
The New York Botanical Garden

Amazonian forests harbor numerous plants of economic value (Balick, 1985; Farnsworth, 1984). In fact, intact Amazonian forests are often more valuable for their timber and nontimber products than are the agricultural plantations or pastures that too frequently replace them (Peters et al., 1989; Menezes, 1990). Moreover, Amazonian forests have inestimable value as reservoirs of biodiversity, regulators of hydrological cycles, protectors of fragile soils, and stabilizers of the atmosphere. Because of the value of Amazonian forests, those who wish to replace them with agriculture or pastures should be required to demonstrate that their projects will yield more than the value of the intact forests.

One of the most important economic plants of the Amazon is the Brazil nut (*Bertholletia excelsa,* family Lecythidaceae). The edible seeds of this species, along with the latex of *Hevea brasiliensis,* are often cited as the most important products of extractive reserves in Amazonia. Brazil nuts are collected mostly during the wet season and rubber is tapped mostly during the dry season. The combination of these two forest products provides year-round income for those living by extractivism.

Collection of the Brazil nuts and rubber has relatively little impact on the ecology of Amazonian forests. Therefore, it is often stated that conservation of biodiversity and exploitation for these and other nontimber products is compatible. However, those who gather nontimber forest products are almost always involved in other activities such as slash-and-burn agriculture, timber extraction, mining, and hunting. As a result, heavily used extraction reserves usually protect only part of the vast biodiversity found in Amazonian forests. Consequently, any conservation plan for the Amazon must include large reserves of all Amazonian ecosystems that are protected from excessive economic exploitation.

In this chapter, I review the natural history, the value of the Brazil nut harvest, the possibilities for plantation cultivation, and the future of the Brazil nut industry. Because of its economic importance, the Brazil nut has been the target of many studies of its biology and agronomy. The greatest number of these studies has been carried out under the auspices of the "Centro de Pesquisa Agropecuária do Trópico Umido" (CPATU) of the "Empresa Brasileira de Pesquisa Agropecuária" (EMBRAPA) at Belém, Pará, Brazil. A bibliography of the Brazil nut with 259 titles is available (Vaz Pereira and Lima Costa, 1981), and a recent summary of Brazil nut biology and agronomy can be found in Mori and Prance (1990b).

NATURAL HISTORY

Taxonomy and Distribution

Bertholletia excelsa belongs to a pantropical family of trees (Lecythidaceae) that includes approximately 200 species in the Neotropics, distributed from southern Mexico into southern Brazil (Mori and Prance, 1990a; Prance and Mori, 1979). The Brazil nut represents a single species in the well defined genus Bertholletia. Although there is considerable variation in fruit size and shape and number of seeds per fruit, there is no justification for recognizing more than one species of Bertholletia.

The closest relatives of *B. excelsa* are among species of Lecythis commonly referred to in the vernacular as the jarana group (Mori and Prance, 1990b). Other species with edible seeds in the family are *Lecythis pisonis* and its relatives, *L. minor* and *L. ollaria*. The latter two species, however, sometimes cause hair and fingernail loss because of excess selenium accumulation in the seeds (Dickson, 1969; Kerdel-Vegas, 1966). Nevertheless, the differences between Bertholletia and Lecythis are so great that there is little hope for introducing genetic material from one genus into the other via hybridization. Consequently, germ plasm for improvement of Brazil nut production will have to come from the variation found within *B. excelsa,* not from closely related species in other genera.

Bertholletia excelsa is an Amazonian plant that prefers nonflooded forest (terra firma) in the Guianas, Colombia, Venezuela, Peru, Bolivia, and Brazil. The climatic conditions under which Brazil nuts grow are summarized in Almeida (1963), Diniz and Bastos (1974), and Mori and Prance (1990b).

Brazil nuts are cultivated in tropical botanical gardens far outside its native range, and minor plantations have been established in Kuala Lumpur in Malaysia (Müller, 1981) and Ghana in Africa (D. K. Abbiw, pers. comm.).

Trees of *Bertholletia excelsa* occur in stands of 50 to 100 individuals that are known as "manchales" in Peru (Sánchez, 1973) and "castanhais" in Brazil (Dias, 1959). Density of Brazil nut trees per hectare varies considerably throughout the Amazon. In a study of Brazil nut production in eastern Brazil, Miller (1990) found from 9 to 26 reproductive trees per hectare, while Becker and Mori (unpublished data) found only one tree over 10 centimeters dbh in a 100-hectare plot in central Amazonian Brazil.

There is some evidence that Brazil nut trees are gap dependent (Mori and Prance, 1990b). Moreover, some authors have suggested that stands of the Brazil nut owe their origin to pre-Colombian Indians (Miller, 1990; Mori and Prance, 1990b; Müller et al., 1980). An understanding of the development of reproductive individuals from seed is still needed before management of Brazil nuts in natural stands is possible.

PHENOLOGY

Flowering of *Bertholletia excelsa* occurs during the dry season and into the wet season. In fact, Brazil nuts grow naturally only in regions with a three- to five-month-long dry season (Müller, 1981). In the eastern part of Amazonian Brazil, flowering begins at the end of the rainy season in September and extends to February. Peak flowering occurs in October, November, and December (Moritz, 1984).

Toward the end of the rainy season, generally in July, the leaves of Brazil nut trees begin to fall. The new growth flushes from directly below the inflorescences of the previous year, and the new inflorescences are produced at the apex of the current growth flush. Large numbers of flowers are produced daily over a relatively long period. The flowers open between 4:30 P.M. and 5:00 P.M. However, the anthers start to dehisce within the bud several hours before the flowers open. The petals and androecia fall in the afternoon of the day that the flowers open (Mori and Prance, 1990b).

Fruit development takes longer in *B. excelsa* than in any other species of Lecythidaceae. Moritz (1984) states that 15 months are needed for the fruits to develop after they have been set. Consequently, Brazil nut fruits fall mostly in January and February, during the rainy season. Under natural conditions, the seeds take 12 to 18 months to germinate (Müller, 1981).

POLLINATION BIOLOGY

The flowers of the Brazil nut are zygomorphic, with an androecium that is prolonged on one side into a hood that arches over and is tightly appressed to the summit of the ovary. In addition, the petals are appressed to the androecium (Fig. 27-1). Consequently, the flowers can only be entered by large-bodied bees with enough strength to pry open the androecial hood to obtain the pollinator reward that is thought to be nectar produced at the apex of the coiled androecial hood. Bees of the genera *Bombus, Centris, Epicharis, Eulaema,* and *Xylocopa* have been captured visiting Brazil nut trees (Moritz, 1984; Müller et al., 1980; Nelson et al., 1985). These bees are nonsocial or semi-social and therefore do not lend themselves easily to manipulation by humans, such as is the case with the social bees (for example, *Apis, Melipona,* and *Trigona*) that can be used to pollinate certain crops by transporting beehives from one plantation to the next.

For the most part, cross pollination is needed for seed set in Neotropical Lecythidaceae. Therefore, the bees, and to a lesser extent bats, are essential for the pollination and subsequent fruit and seed development of Lecythidaceae. Although a low level of in-breeding may occur in *Bertholletia excelsa,* most seed set in this species is the result of cross pollination (Mori and Prance, 1990b). The development of self-compatible lines of the Brazil nut would facilitate plantation cultivation of this species by eliminating the need for cross pollination by the difficult to manage bee pollinators.

Bees outside of the native range of the Brazil nut can effect pollination. For example, Brazil nut trees in Ceylon (Macmillan, 1935), Kuala Lumpur, and Ghana set fruit. However, it is not known if pollinators outside of the native range of the Brazil nut or "weedy" pollinators found in secondary forests are efficient enough to allow for economically viable fruit production. These topics are being addressed by Enrique Ortiz, a Peruvian graduate student studying at Princeton University.

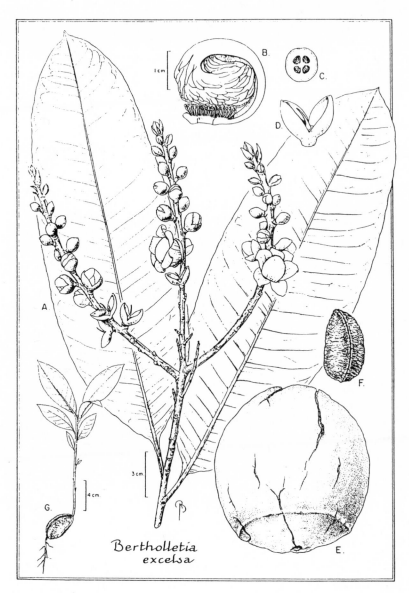

Bertholletia
excelsa

FIGURE 27-1 *Bertholletia excelsa,* A-D (Mori, et al. 17503), E-F, (Prance et al. 16599), G (Prance blastogeny collection no. 4). A. Habit. Note how the petals are tightly appressed to the androecium. B. Medial section of the androecium. Note how the hood appendages are swept inward but do not form a complete coil. Nectar is produced at the base of these appendages. C. Cross section of the ovary. D. Ovary and calyx. Note the 2-parted calyx, a unique feature of this genus. The long, geniculate style points to the open end of the flower where it receives pollen when bees enter the flower but not when they depart.

DISPERSAL BIOLOGY

The fruits and seeds of *B. excelsa* are unique in the Lecythidaceae. At maturity the large, round, woody fruits fall to the ground with the seeds inside. The seeds, which have a bony testa, are removed from the capsules and dispersed by rodents, especially agoutis (*Dasyprocta* spp.) (Huber, 1910; Miller, 1990) and squirrels (Miller, 1990; E. Ortiz, pers. comm.). Agoutis and squirrels may be the only animals able to efficiently gnaw through the extremely woody pericarps. They eat some of the seeds and cache others for subsequent use. Consequently, the seeds are either eaten and destroyed or they are left in a forgotten cache where they may eventually germinate 12 to 18 months later.

BRAZIL NUT HARVEST

Brazil nuts are harvested almost entirely from wild trees during a five to six month period in the rainy season. The fruits, which weigh from 0.5 to 2.5 kilograms and contain 10 to 25 seeds, are gathered immediately after they fall in order to minimize insect and fungal attack of the seeds, and to control the number of seeds carried away by animals (Mori and Prance, 1990b). According to Miller (1990), the number of capsules produced per tree ranges from 63 to 216.

More detailed descriptions of the methods of Brazil nut harvest can be found in Almeida (1963), Mori and Prance (1990b), and Souza (1963).

Collection of Brazil nuts has a major impact on local Amazonian economies. Available figures, however, only provide approximations of total production because of the difficulty in obtaining accurate data from the Amazon. Brazilian production has ranged from 3,557 tons in 1944 to 104,487 tons in 1970. Since 1980, annual production has been around 40,000 tons (Mori and Prance, 1990b). In the past, the welfare of many Amazonian towns, such as Puerto Maldonado, Peru (Sanchez, 1973) and Marabá, Brazil (Dias, 1959) depended heavily on Brazil nut production. In 1986, the total value of shelled and unshelled Brazil nut seeds exported from Manaus alone was $5,773,228. (Mori and Prance, 1990b). Most of the seeds are sent to England, France, the United States, and Germany.

Calculations by Miller (1990) have estimated the primary value (money paid to the collectors) of Brazil nut stands to be $97 per hectare.

This value includes an arbitrary 25 percent discount to allow for seeds left in the stands. The secondary value—in other words, the money received by the exporting company for unshelled nuts by a United States based importing company—was estimated at $175.56 per hectare. The tertiary value—the cost of a bag of unshelled nuts in a Florida supermarket—was calculated at $1059.44 per hectare. Shelled and processed nuts are much more valuable. Efforts, such as those by Cultural Survival, to place some of the shelling and some of the processing in the communities of the collectors provide additional incentive for maintaining extractive reserves. Over a ten year period, utilization of a forest for Brazil nut production appears to be more profitable than extracting timber or cutting the forest for pasture (Miller, 1990).

BRAZIL NUT PLANTATIONS

The methodology for growing Brazil nuts in large plantations has been developed by Müller and his associates of CPATU, Belém, Brazil. An English summary of their work is provided by Mori and Prance (1990b).

As mentioned previously, most Brazil nut production is still gathered from wild trees. However, plantations are being developed in various parts of the Amazon. In January 1990, I visited Fazenda Aruanã, a Brazil nut plantation located at kilometer 215 on the Manaus/Itacoatiara Road in Amazonas State, Brazil. This is a 12,000 hectare former cattle ranch partially converted to Brazil nut plantation in 1980. At the time of my visit, 318,660 Brazil nut trees were planted on 3,341 hectares. In addition, there were an unknown number of individuals growing in nearby forests belonging to Fazenda Aruanã.

The original intent at Fazenda Aruanã was to plant Brazil nut trees in 20 by 20 meter grids and allow cattle grazing between the trees. At first the cattle did not eat the leaves of the trees, but later they did. As a result, the spacing was reduced to 10 by 10 meter grids and the number of cattle was reduced to 300 head.

The trees in the Aruanã Plantation are the result of grafting high yielding clones from the region of Abufari, Amazonas where Brazil nuts are known for their large fruits and seeds. An important consideration in the establishment of this and other plantations is the provenance of the seed. Moritz (1984) has shown that fruit production as the result of fertilization between trees of the same clone is low. Based on this concept, Müller (1981) recommends that buds for grafting should be obtained from at least five different trees. A danger in using so few clones for grafting is that selecting for high fruit yield may also reduce the plantation's ability to resist future attacks of disease and insects.

Root stock is grown on Fazenda Aruanã from seed. The seeds are germinated by softening them in moist sand and subsequently removing the embryo by opening the seed coat along its edges. The embryos, which are treated with a fungicide, germinate in approximately 20 days and the seedlings are grown in plastic bags or cups. When the seedlings are transplanted to the field, about 200 grams of phosphorous is added to the hole in which they are planted. Root stock is ready for grafting at about 1.5 to 2 years when the young trees have become well established in the field. Better scions are obtained if the leaves are removed from the branches 8 to 10 days before the scion is removed. This promotes the formation of vigorous buds that take better to grafting. Thirty days after the graft has been made, the root stock is ringed distally to the graft. This allows the scion to gradually outgrow the branches of the root stock.

After six years, the trees begin to produce fruit. However, at the time of my visit all fruit produced until then had been used to grow more root stock or to produce seedlings for sale to others.

The plantation does not appear to have problems resulting from the lack of pollinators. *Bixa orellana* had been planted to provide pollen as a food source for pollinators. Moreover, there is extensive natural forest surrounding the plantation and windrows with natural vegetation transect the plantation. The windrows harbor numerous plants known to be food sources for the pollinators of Brazil nut trees. Although the trees were nearly 10 years old at the time, there was no evidence or history of disease.

The owners of Fazenda Aruanã have great expectations for this plantation. At the end of 12 years, they expect yields of 4.8 kilograms per tree and $480 per hectare. At the end of 18 years, they predict 8.5 kilograms per tree and $850 per hectare.

FUTURE OF BRAZIL NUT PRODUCTION

Until now Brazil nuts have been gathered mostly from wild trees. In recent years, Brazil nut production has declined because of deforestation, the exodus of Brazil nut gatherers to large metropolitan centers, the flooding of some traditional Brazil nut stands, and perhaps because of disruption of pollinators caused by fires during the dry season when

Brazil nut trees are in flower (Kitamura and Müller, 1984; Mori and Prance, 1990b). If traditional methods of Brazil nut production are to be maintained, very large extractive reserves will have to be established in areas of high Brazil nut tree density.

However, those interested in the preservation of tropical nature should be careful not to equate the establishment of extractive reserves with the maintenance of Amazonian biodiversity. Because Brazil nut gathers and rubber tappers do more than just gather Brazil nuts, they often have a negative impact on plant and animal diversity. Indeed, extractive reserves may become little more than secondary vegetation with economic plants such as Brazil nut and rubber trees scattered here and there. Therefore, the establishment of extractive reserves does not negate the need for well planned biological reserves.

The future success of Brazil nut plantations is still open to debate. Until now, there have been no examples of economically successful plantations of this Amazonian tree. All attempts at growing tropical trees in plantations that do not naturally grow in nearly monotypic stands have been failures. In their native habitat, Brazil nut trees are distributed more or less scattered in the forest in much the same way that rubber trees grow. The economic disaster of attempting to bring rubber into plantations within its home range has been well documented (Hecht and Cockburn, 1989), and there is no reason to believe that Brazil nut plantations in Amazonia will not meet the same fate as the Amazonian rubber plantations. Careful observation of the Aruanã Plantation over the next decade may allow us to determine if Brazil nut production in plantations is a viable alternative to collection from wild trees. If plantations are viable, then conservationists will have to be prepared to assess the impact that plantations will have on the maintenance of extractive reserves.

Finally, it is important that we do not place too much hope on Brazil nut extraction as an economically viable way to support an ever increasing population in Amazonia. In the first place, world markets may not be able to handle much of an increase in Brazil nut production, and, in the second place, such low intensity use of land is not capable of supporting human populations at the level needed to increase the standard of living demanded by more and more people. If Ewel's (1991) estimates that hunting-gathering and shifting agriculture can only support 1 person per 5 square kilometers and 1 person per square kilometer, respectively, are correct (there is no reason to believe that they are not), then extractive reserves will do little to absorb population growth in Amazonian countries. The future of extractive reserves and humanity's ability to preserve a representative sample of Neotropical biodiversity depends on the success of controlling population growth and consumption both inside and outside of the tropics (Erhlich and Erhlich, 1990).

References

de Almeida, C. P. 1963. Castanha do Para: Sua exportação e importância na economia Amazonica. *Edições S.I.A. Estudos Brasileiros* 19:1–86.

Balick, M. J. 1985. Useful Plants of Amazonia: A Resource of Global Importance. In G. T. Prance and T. E. Lovejoy (eds.), *Amazonia*. New York: Pergamon Press.

Dias, C. V. 1959. Aspectos geográficos do comércio da castanha no médio Tocantins. *Revista Brasil. Geogr.* 21(4):77–91.

Dickson, J. D. 1969. Notes on Hair and Nail Loss after Ingesting Sapucaia Nuts (*Lecythis elliptica*). *Econ. Bot.* 23:133–34.

de Diniz, T. D., A. S. Diniz, and T. X. Bastos. 1974. Contribuição ao conhecimento do clima tipico da castanha do Brasil. *Bol. Técn. IPEAN* 64:59–71.

Ewel, J. 1991. Conservation and Agriculture. *Tropinet* 2(1):1.

Ehrlich, A. H., and P. R. Ehrlich. 1990. Extinction: Life in Peril. In S. Head and R. Heinzman (eds.), *Lessons of the Rainforest*. San Francisco: Sierra Club Books.

Farnsworth, N. R. 1984. How Can the Well Be Dry When It Is Filled with Water? *Econ. Bot.* 38:4–13.

Hecht, S., and A. Cockburn. 1989. *The Fate of the Forest*. New York: Verso.

Huber, J. 1910. Mattas e madeiras amazônicas. *Bol. Mus. Paraense Hist. Nat.* 6:91–225.

Kerdel-Vegas, F. 1966. The Depilatory and Cytotoxic Action of Coco De Mono (*Lecythis ollaria*) and Its Relationship to Chronic Seleniosis. *Econ. Bot.* 20:187–95.

Kitamura, P. C., and C. H. Müller. 1984. Castanhais natives de Marabá-PA: Fatores de depredação e bases para a sua preservação. EMBRAPA, Centro de Pesquisa Agropecuária do Trópico Umido. *Documentos* 30:1–32.

Macmillan, H. F. 1935. *Tropical Planting and Gardening with Special Reference to Ceylon*. London: Macmillan.

Menezes, M. A. 1990. Reservas extrativistas: Por uma reforma agrária ecológica. *Ciência Hoje* 11(64):4–6.

Miller, C. 1990. *Natural History, Economic Botany, and Germplasm Conservation of the Brazil Nut Tree (*Bertholletia excelsa Humb. *and* Bonpl*)*. Masters thesis presented at the University of Florida.

Mori, S. A., and G. T. Prance. 1990a. Lecythidaceae—Part II. The Zygomorphic-Flowered New World Genera (*Bertholletia, Corythophora, Couratari, Couroupita, Eschweilera,* and *Lecythis*). *Fl. Neotrop. Monogr.* 21(II):1–376.

Mori, S. A., and G. T. Prance. 1990b. Taxonomy, Ecology, and Economic Botany of the Brazil nut (*Bertholletia excelsa* Humb. and Bonpl.: Lecythidaceae). *Adv. Econ. Bot.* 8:130–50.

Moritz, A. 1984. Estudos biológicos da castanha-do-Brasil (*Bertholletia excelsa*

H.B.K.). EMBRAPA, Centro de Pesquisa Agropecuária do Trópico Umido. *Documentos* 29:1–82.

Müller, C. H. 1981. Castanha-do-Brasil; estudos agronômicos. EMBRAPA, Centro de Pesquisa Agropecuária do Trópico Umido. *Documentos* 1:1–25.

Müller, C. H., I. A. Rodrigues, A. A. Müller, and N. R. M. Müller. 1980. Castanha-do-Brasil. Resultados de pesquisa. EMBRAPA, Centro de Pesquisa Agropecuária do Trópico Umido. Miscelânea 2:1–25.

Nelson, B. W., M. L. Absy, E. M. Barbosa, and G. T. Prance. 1985. Observations on Flower Visitors to *Bertholletia excelsa* H.B.K. and *Couratari tenuicarpa* A. C. Sm. (Lecythidaceae). *Acta Amazônica* 15(1/2):225–34.

Peters, C. M., A. H. Gentry, and R. O. Mendelsohn. 1989. Valuation of an Amazonian Rainforest. *Nature* 339:655–56.

Prance, G. T., and S. A. Mori. 1979. Lecythidaceae—Part I. The Actinomorphic-Flowered New World Lecythidaceae (*Asteranthos, Gustavia, Grias, Allantoma,* and *Cariniana*). *Fl. Neotrop. Monogr.* 21(I):1–270.

Sánchez, 1973. *Explotación y comercialización de la castaña en Madre de Dios.* Ministerio de Agriculture, Dirección General de Forestal y Caza, Informe No. 30. Lima, Peru.

Souza, A. H. 1963. Castanha do Pará: Estudo botánico, químico e tecnológico. *Edições S.I.A., Estudos Técnicos* 23:1–69.

Vaz Pereira, I. C., and S. L. Lima Costa. 1981. *Bibliografia de Castanha-do-Pará* (Bertholletia excelsa H.B.K.). EMBRAPA, Centro de Pesquisa Agropecuária do Trópico Umido. Belém, Pará.

28

Trouble in Paradise: Practical Obstacles to Nontimber Forestry in Latin America

---

Linwood H. Pendelton
Harvard University, Kennedy School of Government

The urgency of rapid tropical forest degradation has achieved a level of awareness that rivals any environmental hazard known to date. The sustainable collection of nontimber forest products may offer a positive step forward in the effort to thwart runaway deforestation. Along with shifting agriculture, nontimber forestry has long been an important means of securing nutrients for many traditional forest peoples (Dunn, 1975; Clay, 1988; Prance, 1989; Cannell, 1989; Eden, 1990). Present-day nontimber forestry relies on the sustainable harvesting of naturally available nontimber forest products (NTFPs) as an economically viable alternative to other destructive agricultural and forestry practices. Nontimber forestry combines the indigenous harvesting wisdom of forest people, the market know-how of business and economics, the attention to biological sustainability demanded by ecologists and agronomists, and the protection of biodiversity demanded by all. In spite of broad academic, political, and popular support, the application of economically and biologically sustainable nontimber forestry faces serious practical obstacles.

The economically beneficial application of nontimber forestry can only be realized within a fairly restricted scenario. The complexity, variety, and reliability required of the elements of this scenario must be considered before every nontimber forestry endeavor. Any successful and sustainable nontimber forestry venture must meet the following prerequisites: 1) there must be a substantial amount (density) of economically valuable nontimber forest products in the area to be harvested, 2) there must be a market for these products and their derivatives, 3) the collection of these products must be profitable in both the short and long run, 4) the net present value of nontimber forestry must be at least as great as the opportunity cost of other forest uses, 5) these resources must be accessible, 6) these resources must be reliably available, 7) these resources must be secure (defensible) over time, presumably through land tenure, and 8) the collection of these resources must be biologically sustainable.

This chapter addresses certain economic prerequisites of nontimber forestry and discusses some prospects and applications of sustainable nontimber forestry in Latin America.

MARKETING NONTIMBER FOREST PRODUCTS

Local Markets

The market dynamics for many nontimber forest products are not well known. Due to perishability problems and strict import regulations in developed countries, most forest fruits can be sold only in local or domestic markets that could be flooded quickly with locally abundant fruits. If local market price elasticities of demand are low, increases in supply will drive prices down. When substitutes are abundant, high cross-price elasticities of demand will encourage substitution. Many potential substitutes may be found in local markets. Vasquez and Gentry (1989) list 43 sweet pulpy fruits that are eaten in the Iquitos area, of which 22 are traded in local markets. The prices for seven of these fruits are given; all prices are comparable. The abundance of substitutes not only keeps market prices low through competition, but even moderate price increases for a single species, due to increased collection costs, could cause a collapse in local demand through substitution.

The Effects of a Strong Local Demand

Local demand may fuel the conversion from forest collection to cultivation. Vasquez and Gentry (1989) show that 18 species of fruit that are collected wild are also cultivated. The fact that the Iquitos area is an important center of diversity for domesticated and semidomesticated fruit trees (Clements, 1989; *see also* Posey, 1985) may be linked to this phenomenon of transitional cultivation. Further evidence of transitional cultivation comes from Brazil; Posey (Clay, 1988) reports that the Kayapo have a long list of cultigens that they claim are of aboriginal stocks. If one forest product proves to be more profitable than others or if comparative advantage is established for one or a few products, then small holders will tend to specialize in these products. When possible, cultivation may be the next logical step in the exploitation of these forest resources.

Even when cultivation is not possible, specialization of one product may be at the expense of more traditional agriculture and subsistence. The strong demand for rubber led many rubber tappers to forsake subsistence agriculture for intensive rubber collection (Murphy and Murphy, 1985). The resulting situation leaves small holders more vulnerable to risk. The same pattern of specialization at the expense of traditional agriculture has also been associated with coca cultivation (Eden, 1990).

International Markets

Strong international markets for potentially sustainable Latin American NTFPs have been established only for a few products. Brazil nuts, certain oils, allspice, coca, and ecotourism indeed are being demanded by international consumers (Clay, 1990; Prance, 1989; Schwartzman, 1989). The markets for these products are not well known. NTFP export prices could fall sharply if 1) new sources are introduced to the international market, 2) synthetic or agricultural substitutes are developed, or 3) changes in consumer tastes occur (Panayatou, 1989). If these products have high cross-price elasticities for demand, they may be vulnerable to substitution (for example, Macadamia nuts for Brazil nuts, beach tourism for ecotourism, synthetic oils for natural oils, plastic for tagua, and so on). If initial demand is based on novelty, the stability of future markets may be tenuous. For novelty items, the risk of substitution may be great. Historically, strong and sustained demand for NTFPs has led to cultivation and domestication (rubber, cacao, coffee, palm oil, cashews, bananas), depletion of wild resources (rattan, fruits, orchids), or synthetic substitution (rubber, palm oils, vanilla, maple).

Collector Profits

For small holders and forest inhabitants to adopt nontimber forestry, NTFP exploitation must be profitable at the collector's level. Unless cartels are formed, "forest-gate" prices probably will be out of the control of collectors. Inputs, however, can be controlled by collectors and the most important input for any NTFP is labor. Most NTFPs occur in low densities throughout the forest (except for the oligarchic forests [Peters, et al. 1989b]), and many literally are out of arm's reach. The marginal product of labor (MPL) is likely to be low for any one product, even though the MPL for multi-product collection could be high. This suggests that the collection of a single NTFP may not be economically lucrative unless it is part of a larger suite of NTFPs. Furthermore, sustainable extraction may be even more labor intensive than unsustainable harvesting (for example, picking fruits instead of felling trees for fruit) and could reduce further the MPL for individual products. Potentially low MPLs for single products could have the following effects: 1) initiation of NTFP extraction may never begin if only one or a few products are targeted for sale, and 2) low MPLs, high discount rates, and insecure land tenure may lead to collecting shortcuts (felling trees, removing all fruit, and so on) and thus nonsustainable harvesting. Local substitutes abound for most NTFPs. If sustainable harvesting results in increased costs and prices, sustainably harvested NTFPs are susceptible to substitution by cheaper, nonsustainably harvested competitors. Unless a diversified and balanced portfolio of NTFPs are collected simultaneously, collectors will be vulnerable to even modest market fluctuations.

PRODUCT ACCESSIBILITY AND AVAILABILITY

Accessibility

For economically sustainable nontimber forestry, nontimber forest products must be accessible and reliably available. Getting the product to market—or in the case of ecotourism, getting the market to the product—may prove prohibitively expensive or even logistically impossible. Market accessibility requires a minimum of infrastructure. In the tropics, experience has shown that the establishment of infrastructure (roads) usually goes hand-in-hand with accelerated colonization and thus forest destruction (Eden, 1990). Forest destruction and nontimber forestry tend to be incompatible. Even when infrastructure is available, high transportation costs may render nontimber forest products uncompetitive relative to other substitutes. Due to poor infrastructureal links within the Autonomous Regions of Nicaragua, imported rattan is cheaper than its locally available close substitute bejuco de la mujer (Pendelton, work in progress).

Availability and Risk

Infrastructure aside, many nontimber forest products may not be available on a reliable basis. Natural competition can be fierce for fruits and seeds in tropical forests. Some suspect that little marketable fruit is left after birds and primates have finished with the unripe and ripening fruits. Weather also plays an uncontrollable role in resource abundance. Little can be done to protect resources against weather hazards. Replanting is simply not an option.

Risk averse small holders may be unlikely to undertake sustainable nontimber forestry. Unpredictable climatic variations can change drastically the abundance of available products. Unlike irrigated cultivation, climatic and environmental factor-inputs cannot be controlled. For nontimber forestry, water needs are met by rainfall; nutrient needs come from complex forest nutrient cycles; traditional competitive enhancement has been explored only recently (Anderson and Posey, 1985; Anderson, 1990). Left to the vagaries of nature, the variance of nontimber product yields is likely to be high. A high yield variance, little or no rural credit or insurance, and unproven markets could further increase the perceived opportunity costs of nontimber forestry.

LAND TENURE

A near universal lack of serious land tenure in tropical forested countries brings into question the broad applicability of nontimber forestry. Most experts agree that harvesting in perpetuity, or at least into the future, must be guaranteed for nontimber forestry to be economically more valuable than other destructive forest uses. Following Peters's calculations (Peters et al., 1989a), as the number of annual harvests decreases, so too does the net present value of nontimber forestry. Alternative forest uses, especially logging and ranching, are far less sensitive to decreasing length of land possession. For logging, ranching, and sometimes agriculture, the rewards from clearing the forest are essentially immediate. Without land tenure, or a suitable alternative, it may be in the best interest of forest people or colonists to convert their resources as quickly and as lucratively as possible. The importance of land tenure cannot be overstated. Of all the variables involved in the equations of successful nontimber forestry, any land tenure can be manipulated directly through public policy.

PROSPECTS FOR NONTIMBER FORESTRY

There are tropical forest localities where the myriad of elements needed for lucrative nontimber forestry may be found. These proceedings catalog what may be the most promising possibilities for sustainable nontimber forestry endeavors. The commercialization of NTFPs is already underway in the Petén of Guatemala, in Ecuador with the Tagua Initiative, and throughout Brazil and Peru with Brazil nuts. Yet not one of these countries is insulated from the economic ravages that have destroyed Latin American economies, governments, and entrepreneurs.

Why and Where Nontimber Forestry?

WHY: THE PURPOSE OF NONTIMBER FORESTRY. Before nontimber forestry can be prescribed as a forest use, the goal of nontimber forestry must be made explicit. This chapter has treated nontimber forestry as one of many techniques that may help forest-dwelling small holders. In the absence of active governmental or extra-governmental intervention, the small holder perspective is the one most likely to predict the ultimate choice of forest use.

OPTIMAL PATHS OF FOREST DEVELOPMENT. In a static sense, nontimber forestry may meet the simultaneous goals of improving small holder welfare and improving social welfare. From a dynamic perspective, however, the optimal paths for the continued improvement of small holder welfare and social welfare may diverge. In the case of extractive reserves, resident collectors are prohibited from clearing large areas of forest for forest product cultivation. This may be sound policy for maintaining or increasing social welfare by protecting the forest and the services it provides. Nevertheless, prohibiting NTFP cultivation may hamper the continued improvement in small holder welfare, especially when cultivation is economically superior to collection or if cultivation is imminent in areas outside the reserve. When it is the case that the path of optimal social welfare improvement differs from the path of optimal small holder welfare improvement, small holders should be compensated if forced to forgo the enhanced welfare opportunities of forbidden forest development.

WHERE: COMPARTMENTALIZATION. Rain forests in the Neotropics vary considerably both between forests and within forests. As a result, no single forest use is likely to be suitable for all forest types. Wise forest management demands the daunting risk of matching forest use techniques to appropriate forest types. Although difficult, the recognition of forest types and the differentiation of forest uses is possible. In fact, ecological and forest-use zoning has long been part of the Kayapo way of life (Posey, 1983). Odum (1969) and later Eden (1978) recognized the need for forest development zoning. Both prescribe a system of compartmentalization in which different land types are devoted to different but optimal land uses. For many traditional peoples, this means agriculture in the fertile varzea forests with collecting in nearby terra firme forest. That the two activities are pursued simultaneously may signal that either activity alone might not be sustainable or adequate.

The Wrong Forests

HIGH SPECIES DIVERSITY FORESTS. Nontimber forestry should not be initiated in forests of exceptionally high species diversity. Forest collectors depend on a number of forest-altering activities in order to meet basic subsistence needs. Most important among these activities is land clearing for agriculture, hunting, and forest management (enhancement). While land clearing can be contained locally, hunting tends to follow existing trails. When trails are extensive, as they must be for nontimber forestry, hunting can lead to wildlife depletion deep into the forest (Wilkie, 1990). To improve the economic returns to nontimber forest products, collectors will attempt to reduce collecting costs through forest management. For some forests, this means clearing away vines to get at fruits or opening the understory to encourage the growth of herbs. In the extreme, it may mean exterminating animal competitors and potential pests. Regardless of the method, species diversity in managed forests tends to be lower than that in unmanaged forests (Anderson, 1990). When protecting species diversity is the end, nontimber forestry should not be the means.

UNCOLONIZED FORESTS. Nontimber forestry should not be encouraged in forests in which there is only minimal social organization. Nontimber forestry requires infrastructure and a highly organized social network. The dangers of infrastructure development are well known and include spontaneous colonization and wide-scale forest destruction. Nontimber forestry carries with it inherent costs, including deforestation, species depletion, and potential conflicts with or among indigenous peoples. Unless forest colonization is imminent, remote forests will be better left unvisited.

WETLAND FORESTS. Wetland forests traditionally have been the sites of agriculture and villages (Eden, 1990). Relatively fertile, flat soils will almost certainly give higher returns to agriculture than nontimber forestry. Furthermore, wetlands traditionally have supported much higher population densities than other forests (Eden, 1990). Exploiting the high carrying capacity of wetland forests will help to alleviate population pressures on more internal forest areas. (Traditionally, the terra firme forests have sustained population densities two orders below that of wetland forest [Eden, 1990]). From the point of view of macroeconomics, nontimber forestry is probably not the optimal use of wetland forests. From a conservation perspective, the less species-diverse wetlands (Eden, 1990) may provide an important and sustainable safety valve for the protection of more diverse terra firme forests.

The Right Forests

FOREST MARGINS. In most cases, nontimber forestry probably will never stand alone as a livelihood for small holders in Latin America. Nontimber forestry, by itself, also is unlikely to be economically justified as the best economic use of most tropical forests. Nontimber forestry, however, can indirectly curb deforestation in Latin America by increasing the income of small holders living within the forest and along forest margins. In this sense, nontimber forestry serves to keep small holders from moving deeper into the forest.

BUFFER ZONES. The employment of nontimber forestry in forest margins is perhaps most appropriate in the buffer zones of wilderness parks and biosphere reserves. As long as the core areas of protected forests are well protected or sufficiently remote from nearby settlements, regulated nontimber forestry in buffer zones could provide a source of cash income for neighboring small holders, increase the economic value of the buffer zone, and provide a de facto monitoring of forest activity (for spontaneous colonizers or illegal timbering).

COLONIZED FORESTS. Where land tenure laws only allow for partial clearing of land, small holders could enhance the NTFP potential of forested holdings by leaving intact abutting forests on adjacent plots. These areas could then be managed cooperatively for their NTFPs. Maintaining adjacent forests would 1) enhance species diversity and species survival by increasing the effective size of habitat islands, 2) increase the economic benefits of nontimber forestry by increasing the area harvested, 3) provide habitat for wild game—an important source of protein, and 4) provide peer pressure against cutting down remaining forest. Alone, the forested remnants of one small holder's property may not provide or possess the economic benefits needed to encourage their management. Together, however, agrarian small holders can achieve an economy of scale that may encourage forest fragment management.

WATERSHEDS. Watersheds, and other areas in which environmental externalities are large, are sites in which the social potential for nontimber forestry may be high. In populated watershed areas, nontimber forestry should be encouraged by providing economic incentives that are tied to the implementation of nontimber forestry over other activities. These incentives could include tax breaks, government subsidies (including food subsidies), access to health care for families of collectors, and expedited titling for nontimber forestry concessions.

CONCLUSION

For certain Latin American forests, nontimber forestry may meet the simultaneous goals of improving small holder welfare and protecting social welfare interests by maintaining important forest services. Yet the successful implemenation of nontimber forestry requires a number of prerequisites including an abundance of marketable nontimber forest products, a market for these products, economically and financially profitable returns to NTFP collection, resource accessibility and availability, and basic property rights.

Despite the obvious social benefits of nontimber forestry, the small holder's perspective must be considered when judging the potential of nontimber forestry in Latin America. In most cases, the implementation of a particular forest use cannot be mandated but ultimately is the choice of the small holder. This chapter argues that the risk associated with markets, yield, and property rights may prevent or inhibit certain small holders from adopting nontimber forestry over other forest uses. Even when a static view of nontimber forestry proves economically optimal from the perspectives of both social and small holder welfare, the optimal paths for improving social and small holder welfare may diverge substantially. When this is the case, small holders may abandon nontimber forestry for more profitable endeavors. When continued nontimber forestry is mandated, social welfare may be maintained at the expense of increases in the welfare of small holders.

When all prerequisites are met, nontimber forestry may be an appropriate forest use in forest margins, colonized forests, and forests in which environmental externalities are large. In particular, nontimber forestry appears to be especially applicable to buffer zone forests around protected forest and forests protecting important watersheds. Nontimber forestry is not likely to be the optimal forest use for areas of high biodiversity, uncolonized forests, and wetland forests.

Alone, nontimber forestry is impotent in stopping deforestation by agrarian small holders that continue to practice shifting agriculure. In many cases, the need for cash crops dictates the agricultural techniques employed by small holders. Nontimber forestry, however, may reduce the need for cash crop cultivation by providing small holders with an alternative currency-generating commodity. In this way, nontimber forestry may curb the extent of shifting agriculture and encourage the adoption of more sustainable agricultural methods. With proper incentives, nontimber forestry also can be used to enhance the financial and economic returns to areas in which maintaining a standing forest is the optimal path for improved social welfare.

References

Anderson, A. B. 1990. Extraction and Forest Management by Rural Inhabitants in the Amazon Estuary. In A. Anderson (ed.), *Alternatives to Deforestation: Steps toward Sustainable Use of the Amazon Rain Forest*. New York: Columbia University Press.

Anderson, A. B., and D. A. Posey. 1985. Manejo do campo e cerrado pelo os indios Kayapó. *Boletín do Museo Paraense Emilio Goeldi*. Belém, Brazil: MEPEG, cited by Clay 1988.

Cannell, M. G. R. 1989. Food Crop Potential of Tropical Trees. *Experimental Agriculture* 25:313–26.

Clay, J. W. 1988. Indigenous Peoples and Tropical Forests. *Cultural Survival:* Cambridge, MA.

Clay, J. W. 1990. A Rain Forest Emporium. *Garden,* Garden Society, NY, 14: 2–7.

Clements, C. R. 1989. A Center of Crop Genetic Diversity in Western Amazonia. *Bioscience* 39:624–31.

Dunn, F. L. 1975. Rain Forest Collectors and Traders: A Study of Resource Utilization in Modern and Ancient Malaya. *Monographs of the Malaysian Branch Royal Asiatic Society* (5).

Eden, M. J. 1978. Ecology and Land Development: The Case of Amazonian Rain Forest, *Transactions of Institute of British Geographers*, N.S. 3:444–63.

Eden, M. J. 1990. *Ecology and Land Management in Amazonia*. London: Bellhaven Press.

Murphy, Y., and R. F. Murphy. 1985. *Women of the Forest*. New York: Columbia University Press.

Odum, E. P. 1969. The Strategy of Ecosystem Development, *Science,* 164:262–70.

Panayatou, T. 1989. Modelling Optimal Resource Depletion in Developing Countries: A Joint Investment-Extraction Decision. *Development Discussion Paper No. 286*. HIID, Cambridge, MA.

Pendleton, L. H. (in progress). A Study of the Economic and Ecological Sustainability of Nontimber Forestry in Nicaragua: *Bejuco* Harvesting in the Autonomous Regions.

Peters, C. M., M. Balick, F. Kahn, and A. Anderson. 1989b. Oligarchic Forests of Economic Plants in Amazonia: Utilization and Conservation of an Important Tropical Resource. *Conservation Biology* 3(4):341–49.

Peters, C. M., A. H. Gentry, and R. Mendelsohn. 1989a. Valuation of an Amazonian Rainforest. *Nature* 339:655–56.

Posey, D. A. 1983. Indigenous Ecological Knowledge and Development of the Amazon. In E. Moran (ed.), *The Dilemma of Amazonian Development*. Boulder, CO: Westview Press.

Posey, D. A. 1985. Indigenous Management of Tropical Forest Ecosystems: The Case of The Kayapo Indians of the Brazilian Amazon. *Agroforestry Systems* 3(2):139–58.

Prance, G. T. 1989. Economic Prospects from Tropical Rain Forest Ethnobotany. In J. O. Browder (ed.), *Fragile Lands of Latin America, Strategies for Sustainable Development*. Boulder, CO: Westview Press.

Schwartzman, S. 1989. Extractive Reserves: The Rubber Tappers' Strategy for Sustainable Use of the Amazon. In J. O. Browder (ed.), *Fragile Lands of Latin America, Strategies for Sustainable Development*. Boulder, CO: Westview Press.

Vasquez, R., and A. W. Gentry. 1989. Use and Misuse of Forest-Harvested Fruits in the Iquitos Area. *Conservation Biology* 3(4).

Wilkie, D., and J. G. Sidle. 1990. Social and Environmental Assessment of the Timber Production Capacity of Extension Project U.S.A.I.D./O.I.C.D.

29

The Tagua Initiative in Ecuador: A Community Approach to Tropical Rain Forest Conservation and Development

RODRIGO CALERO HIDALGO
CIDESA, Ecuador

NORTH ESMERALDAS: THE PROJECT AREA

The northwestern province of Esmeraldas in Ecuador is one of the most important regions in the world due to its high biodiversity. For hundreds of years, the forests of Esmeraldas faithfully reflected its name (the greenery of the emerald). However, this special trait has been changing dramatically in recent years.

The tropical rain forest of Esmeraldas presently covers a surface of approximately 1.3 million hectares. It is estimated that in this region there are some 500 plant species, of which perhaps 230 are particularly important; of the latter, some 25 or 30 are utilized, albeit inappropriately.

This biodiversity is threatened by the high prevailing rates of deforestation in the region and throughout the country.[1] Of all the countries in Latin America, Ecuador has the highest rate of deforestation, equal to 2.3 percent annually. At this pace, according to the specialists, Ecuador will become an ecological desert within 40 years.

The province of Esmeraldas covers an area of about 1.5 million hectares, most of which is suited for agricultural and forest activities. Until the decade of the 1950s, this province was known for subsistence agriculture. Nevertheless, its forest wealth linked it to world markets through such products as tagua (vegetable ivory), rubber, and balsa wood. Between 1948 and 1965, banana production was vigorously developed. After the banana market began to decline in 1966, timber harvest became more important. For some agrarian sectors, timber turned out to be more profitable than farming, and thus logging continues unabated. Indeed, it is accelerating.

Since September 1990, the Tagua Initiative (TI) has been implemented in the territory of the Río Santiago Cayapas Commune. The Commune is an agrarian organization of the local black population, with a history of over 105 years. It enjoys legal standing as an entity recognized by the government, and is comprised of 52 member communities that are settled in an area of about 63,000 hectares.[2] Paradoxically, this territory has not yet been formally recognized by the government.

According to recent estimates, the Commune population consists of some 15,000 inhabitants, with an annual population growth rate of 3.7 percent. Four of every ten children suffer some degree of malnutrition; the child mortality rate is at 6 percent. There is widespread incidence of tropical diseases such as malaria, onchocerciasis (river blindness), leshmaniasis, as well as other contagious illnesses such as tuberculosis and leprosy. There is a high rate of alcoholism and drug addiction (with the current danger that the situation with the cocaine cartels of Colombia will convert Esmeraldas into a zone of production). Life expectancy is about 50 years of age and average monthly income per family is under $80 (Comuna Río Santiago-Cayapas, 1990). Another aspect of the substandard living conditions is the high level of cholera in the Commune area, which, along with other problems, is due to the total lack of basic services such as clean drinking water, sewerage, and adequate sanitation and health infrastructure.

The total economic, social, and political exclusion of the Commune from mainstream society is one of the principal reasons for the impoverishment of the population that is described above.

The traditional forms of culture, local organization, and of management and use of natural resources have inevitably been losing out to other patterns introduced and imposed by different means. Some such means, such as those of the religious missions, are nonviolent and slow, yet persistent. Others are less peaceful, such as those of the lumber, palm heart, and banana companies. These are often supported and given the appearance of legitimacy by corrupt government officials.

The struggle by the people of the Commune for their territories, the inviolability of their lands, and, thus, the right to choose their own lifestyles, is now—and perhaps will remain—the main problem of the Commune. It has also become the main objective in the Commune's struggle with the government and mestizo society at large. There have been many battles along these lines, just as many as the attempts to take away the lands. But until now, the people of the Commune have been able to protect their lands. We do not know for how much longer they will be able to protect these lands. Perhaps a good deal of encouragement may be found in the words of the former President of the Commune, Flavio Valdez, who declared to national officials during a meeting in Borbón in November 1990:

> These lands are ours, not only because our forefathers purchased them with their sweat and their labor, but also because on them we were born, and on them we have always lived. This is why these territories have to remain ours as long as we live.

For all of the above reasons, one can conclude that the fundamental role of the Commune is to defend the inviolability of the lands. However, appropriate means to resolve the growing needs of the communities have not been developed. This has been fertile ground for the presence of the companies, brokers, and new settlers who make up the trinity of plunder that preys on the resources of the region.

THE TRADITIONAL SYSTEMS OF MANAGEMENT

The traditional management of the tropical rain forests by indigenous groups has, for decades, provided a relative equilibrium between humans and nature in the northern region of Esmeraldas.

There are two kinds of zones in the area of the Commune: 1) the individual farm, of varying size, delivered by the Commune for its members to use, and on which there may exist different systems of production and extraction, and 2) the communal reserve, which is mainly used for hunting, but where lumber is currently extracted by the Commune members as well as by the Organization itself through its leaders.

There are two or three kinds of usage sites on the territory of any given individual farm. The first is called the "chacra," which is the area of agroforest production, where some 20 to 22 agricultural, forest, and multi-use species are managed, mainly as subsistence crops. Small farm animals are also raised there, and, whenever possible, lumber is extracted.

The second usage site is the reserve where principally forest species are cared for and extracted. Some extracted species include commercial products such as tagua (vegetable ivory), rubber, and the royal palm

(*Attalea colenda*). The reserve is also an area used for hunting. A brief resource survey taken in July 1990 found an average of 18 to 20 usable forest species, as well as 13 edible animals, including fowl.[3] Thus, this area is used to extract such commercial products as lumber and tagua, to hunt, and to cover the basic food needs of the family.

The third kind of usage site is the pasture, which is not found on all farms. Usually, it is an area that has been almost entirely deforested, where grass for cattle grazing is grown and where commercial forest species, such as laurel and cedar in particular, are kept up or planted anew. The species mentioned here are simply those few products used for commercial purposes or family subsistence; they represent a mere fraction of the total biodiversity.

By contrast, during a brief survey undertaken in September 1990 by Conservation International specialists (Drs. R. Foster, M. Plotkin, and H. Pedersen), 330 plant species were identified from 82 distinct botanical families. Among them were 25 tappable forest species, and 13 fruit-bearing species of possible agri-industrial use. Excluded from this survey were the annual plant species that have been introduced, such as rice, pineapple, peanut, kidney bean, and others whose numbers continue to increase (Foster, 1991).

Tagua (*Phytelephas aequatorialis*), which is also known as vegetable ivory, is one of the most common components of the traditional systems of rain forest utilization in the region. In the territory of the Commune, it is distributed sporadically, in patchwork fashion over high hills and in corridors along the flood banks of the rivers Santiago, Cayapas, Onzole, Bogota, and estuaries that flow into them (Foster, 1991).

The tagua palm, particularly those growing along the riverbanks, is associated with an extensive biodiversity of commercial native and exotic trees; in particular, cacao (*Theobroma cacao*), the breadfruit tree (*Artocarpus altifilis*), and the coconut (*Cocos nucifera*) (Bernal and Galeano, 1991). This represents an important starting point from which to build a sustainable mixed agriforest system (Foster, 1991).

Before the TI got under way, the tagua groves were basically abandoned. The situation has begun to change because the Commune members have started to clean out the groves and care for them, albeit in an incipient fashion.

THE COMMUNE'S INTEREST IN THE TAGUA INITIATIVE

The situation prevailing in the communities is one in which activities are carried out amid unstable, poverty-stricken living conditions, with little real prospects for change because neither the government nor the development agencies have responded to the interests of the groups in the communities. Therefore, the inhabitants of the Commune have been forced into a process of indiscriminate use of resources in order to meet their most basic needs. Nevertheless, these grassroots groups have sought alternatives that would enable them to change this situation, handle their resources in a sustainable manner, and improve their living standards. However, they have almost no support from either the government or the NGOs.

It is in this context that the Tagua Initiative (TI) promoted by Conservation International (CI) has been launched. The TI attempts to manage tropical forests in an alternative way by responding simultaneously to the needs of preserving available resources and the needs of the local inhabitants. This requires a development strategy that encourages tropical resource management, use, and preservation for the benefit of the local inhabitants. This is what we in CIDESA call socio-environmental development. An important element of this alternative is the active role of the community throughout the entire process, including the direct administration of resources awarded by donor agencies. The technical support and training, which the NGOs and the government can and must offer, have to encourage empowerment as well as open up more flexible and advantageous terms of trade with the industrialized countries. This is the model for the Tagua Initiative.

The TI is developing in two different settings: one is in the United States, where Conservation International is working to promote and expand the market for tagua and other tropical forest products by stimulating greater appreciation for the renewable natural resources that are managed by the producers themselves. The second setting is in Ecuador, where the Río Santiago-Cayapas Commune and the CIDESA Foundation are cooperating with technical assistance from CI experts.

The pace of implementation in each setting will be different, as will be the difficulties encountered. U.S. businesses are eager to introduce products rapidly while implentation at the community level must proceed slowly, allowing the community to take over the initiative, administer, and develop it. The pace of such a process cannot be imposed. In helping to coordinate these two different paces, CIDESA, as the local NGO, plays an important buffer role.

These were the premises that the Commune and its leadership accepted when they agreed to meet the challenge of implementing the TI in their territory, which began in late September 1990.

An important aspect of the TI is that the resources directed by Conservation International to the Commune will finance more than just the marketing of unprocessed tagua. A significant portion will also go to diversify the supply of this product, through some degree of processing (for instance, shelled tagua, disks, and handcrafted products); likewise, the resources will help improve the sustainable production of other forest and/or agricultural products that the Commune members currently produce (such as cacao and plantain). Furthermore, some of the resources will support the improvement of the administrative capacity of the Commune, including the development of a local promotion team to assume responsibility for the training and basic technical assistance that will be provided to producers.

WHERE ARE WE NOW?

For methodological purposes, the TI is divided into two phases: Phase I has consisted of the commune directly marketing unprocessed tagua on the national market. Phase II involves the development and diversification of tagua marketing and production, as well as conservation and management activities of tropical forest resources. The strategy is to create the most efficient yields and profitable marketing of current products in order to generate greater income for the local residents involved in the extraction, farming, and marketing of forest products. Phase II also involves relieving the pressure on the timber species that at present consitute the quickest means by which resources can be obtained to meet vital needs.

Phase I has now concluded. The most important progress has been in five areas:

1. The Commune has established a local community mechanism to collect and market unprocessed tagua. These tasks are under the purview of a Marketing Committee of five members, of whom one is a CIDESA delegate, and another is a leader of the community council. Both of them engage in assistance and supervision of the business process. Also represented on the Committee are community buyers, who purchase tagua from the Community members in the 12 collection centers located in the largest community. There are two large warehouses where the tagua from the 12 collection centers is stored prior to being shipped to the main factories in the ports of Manta and Guayaquil. The Marketing Committee has a ship with the the capacity to transport 5 tons of tagua along with basic equipment needed for its purchase, such as scales and containers. Young men from one of the communities make up the stevedore team that loads and unloads the tagua coming both

from the collection centers to the warehouses, as well as from the latter to the factories.

2. A simple management accounting system has been established, which is run by the Commune members themselves. It provides the Commune, as well as the donor agencies, with sufficient information about resources for supervising, auditing, and follow-up planning and evaluation. As the entity responsible for managing the purchase and sale of tagua, the Marketing Committee has received basic training in areas of management and accounting from CIDESA. The great advantage of this system is that it can be used to track any kind of financial resources.

3. A practical system has been developed to facilitate relations between producer and consumer (that is, respectively, between the Commune and the clothing and button manufacturers in the United States) (CIDESA-ECUADOR, 1991). Within this system, so long as the Commune itself has not yet begun to produce disks of tagua, the link between the producer and consumer is made via selected national factories. These factories have contracts with CI-authorized button factories in the United States and purchase their tagua from the Commune under the most favorable conditions. As of June 1991, after six and a half months of having its marketing system in operation, the Commune has collected 150 metric tons of tagua, of which approximately 130 metric tons have been sold since March 1990. The remainder is in the process of being dried, prior to being shelled and exported. The market Committee's average monthly collection rate ranges from 20 to 25 tons, which represents 30 to 35 percent of the total amount of tagua produced in the area. The rest is purchased by four brokers who have been active in the area for decades.

4. The price the producer receives for the tagua has risen an average of 27.5 percent since December 1990, although these increases have varied by community (CIDESA-ECUADOR, 1991).

5. Among the Commune members, interest has grown in getting involved in the management activities of the tagua groves and of the other forest species and production systems. This has created a strong foundation on which to implement Phase II of the TI.

The TI has had to face a number of problems in this period: the apparent apathy, with regard to the activities that have to be performed, among a segment of the Commune leadership and many of its members (particularly those who have do not have tagua groves). Moreover, there has been outright opposition and competition among the local brokers who feel that the Commune's direct marketing of tagua endangers their ability to continue cheating the producers from whom they purchase the tagua and that they are threatened with the total loss of power in the region. We have also had to face the opposition of the Catholic religious mission and its NGO, which has worked in the region for almost 20 years.

Perhaps the most serious problem has been the delay among the Commune leaders and members in taking over the TI and making it their own. We understand, however, that this is a normal process, particularly inasmuch as this is the first project run by the Commune's own organization, and the history of its relationship with the government and with other NGOs has been laden with deception and failure.

THE VISIBLE EFFECTS FROM PHASE I

Despite the relatively short time that has transpired, we believe that visible effects can already be seen in different areas. From an organizational perspective, the work methodology and the resource management system utilized have furnished the Commune, its leaders, and a large segment of its members with an ongoing insight into their own talents and limitations and an understanding that in order to perform activities, regardless of how simple they may be, planning and evaluation are necessary. The exercise of these activities has gradually become routine for the leadership and managers. Coincidentally, a small group of Commune members have, since February 1991, been pressuring the Commune to implement a banana plantation project, with the support of an Ecuadoran export firm. Their goal is to plant 12,000 hectares of bananas on Commune lands. This development, the purpose of which is diametrically opposed to the TI, allowed us to learn to what degree the Commune member producers were imbued, if not with the notion of the TI per se, at least with the ideas and alternatives of forest and available resource management. This was a fine opportunity to reflect on the ecological implications of setting up the banana plantations, as well as a chance to consider the social and organizational effects of the proposed project. The prospects of losing their lands and becoming displaced by settlers who would arrive as a cheaper labor force and of losing the diversity of products that their farms yield have led Commune members who initially welcomed the idea of the banana plantation to lose their initial enthusiasm and join the ranks of those who opposed the project from the start. The current prospects make it unlikely that even 25 percent of the originally envisioned project will be carried out. The episode also taught us the lesson that, under the conditions of poverty and deprivation, it is easier to buy the consciences and decisions of the leaders and Commune members, and that the more uprooted from their own communities the leaders become, the more easily they will fall into the traps that are often made for them by economically powerful groups.

The implementation of a marketing system via community collection centers, negotiations over prices with factories in Manta, hiring personnel to load the tagua, and managing bank accounts all help develop leadership and open a new dimension in the relationship between the Commune members and the brokers, the NGOs, and the government itself. Thus, the once unimpeachable brokers are now questioned; the producer feels confident about demanding and even choosing to whom to sell tagua. Thus, the brokers have resorted to a diverse series of manipulations in order to continue obtaining tagua. In fact, as noted above, four large brokers account for 65 percent of the market for the tagua extracted from the area of the Commune. In the last six months of the TI, they have lost a market share that they had maintained for quite some time.

The experience over these months has taught many lessons about the role and power of the brokers. It is possible that the Commune will have to continue to live with the brokers for a long time to come. This may be a positive factor in the sense that the existing competition will always force the organization to operate efficiently. On the other hand, as the Commune becomes more experienced in the marketing process, setting prices and standards of quality, the brokers will be forced to participate in the process, with ultimate results to the benefit of the producer and the management of the resource.

Phase I of the TI has enabled the Commune to learn a lot by negotiating its relationship with the government, other NGOs, and CIDESA itself. We can attest to the fact that a more critical attitude is being developed toward the presence and activities of the agencies. It strikes us as important to encourage and support this attitude to the extent that it permits the Commune greater decision-making autonomy, and it provides a strong foundation on which to achieve empowerment.

In this same process of self-assessment or training and apprenticeship, there have been and will be false steps. This, however, strikes us as a natural indicator that changes are taking place and that new attitudes are developing. The task ahead is to identify and create the conditions that will prevent false steps. Time, patience, perseverance, and a deep commitment to the community, the people, and their futures are required.

CONCLUSION

We believe that the TI can and should serve as an example for new forms of relationships among groups that are involved and interested in environmental conservation in general and tropical forest conservation in particular. Important premises of these new relationships are mutual respect and the reliability of the procedures and interplay of the parties involved. CIDESA believes that the fate of the tropical forest that still remains depends largely on the participation and involvement of the communities that live in it and off it. On the basis of the attitude that they adopt toward the communities dwelling in the tropical forests, the agencies, official and nongovernmental, national and international, can either support conservation or open the way to quicker destruction.

As stated by Dr. Robin B. Foster in his 1990 report, "The Forest and Phytelephas Populations in Comuna Rio Santiago" for Conservation International, the Tagua Initiative "is still an experiment in conservation, but a good one. Even in the worst case it is an excellent investment in promoting the use of natural sustainable resources in the lowland tropics."

An important premise for the conservation of tropical forests is the sustainability of the productive and extractive processes that take place therein. This sustainability is tied to ecological, social, and economic considerations. The three areas are interrelated. However, the profitability of these new forms of forest use, including the TI, are subject to the changing demands of the international market. Therefore, profitability cannot guarantee sustainability.

We understand that the preservation of tropical forests is not only a vital necessity for the inhabitants in these areas, but for humanity itself, because of the direct relationship with the maintenance of biodiversity and thermal equilibrium. As such, it is clear that the industrialized countries, the principal contributors to worldwide pollution, must accept the responsibility for guaranteeing the economic sustainability of the tropical forests; not merely through grants and assistance from private agencies, but also by establishing a demand sustained on the basis of conditions of balanced trade. Who will take on the task of instilling a real and effective consciousness among the consumers?

A last consideration has to do with a certain ambiguity in the attitude of the conservation agencies toward mechanisms to promote conservation in our countries. On the one hand, they are supporting alternative processes through local groups and organizations; on the other, they are supporting the governments that have yet to define or implement clear-cut conservation policies, but rather continue to promote deforestation and environmental pollution. In our judgment, greater prescience is called for in the global attitude of the agencies and donor governments in order to bring what is proposed in line with what is supported.

Notes

1. Data presented at the Meeting of the International Tropical Timber Organization held in Quito from June 1 to 7, 1991 indicated that each year in Ecuador, 350 hectares are deforested, of which only 10 percent is reforested.
2. According to the most recent data of the Ministry of Agriculture and Livestock, about 56,000 hectares of Commune lands are within the State Forest Preserve (June 1991). The Commune is currently appealing for the exclusion of this area from the category of government "protection."
3. This data is taken from the results of a preliminary brief survey carried out during the training seminar for the use of the BESAR methodology, promoted by the World Resources Institute in July 1990. The seminar was undertaken in the area of the Río Santiago Cayapas Commune.

References

Bernal, Rodrigo, and G. Galeano. 1991. *Reporte de Visita al Proyecto Tagua en Ecuador.* Fundación Inguede para la Conservación del Trópico. Instituto de Ciencias Naturales; Universidad Nacional de Colombia, Bogotá.

Calero H. R. 1990. *La Iniciativa Tagua: Una Alternativa de Conservación para el Trópico Húmedo de Ecuador.* Fundación CIDESA., Quito.

CIDESA-COSTA RICA. 1989. *Hacia una Estrategia de Desarrollo Socio-Ambiental. Documento de Trabajo,* San Jose.

CIDESA-ECUADOR. 1991. *Reporte Programático del II trimestre de la Iniciativa Tagua,* Quito.

Comuna Río Santiago-Cayapas. 1990. *Proyecto "Manejo Comunitario y uso sostenido del Bosque húmedo Tropical en las tierras altas de la Provincia de Esmeraldas,"* Quito.

Foster, R. B. 1991. Report: *The Forest and Phytelephas Population in Comuna Río Santiago, Esmeraldas, Ecuador.* Washington, D.C.: Conservation International.

Ministerio de Agricultura and Ganadería. *Subsecretería Forestal y Recursos Naturales (SURFOREN).* Situación Actual del Sector Forestal. In Plan de Acción Forestal. Diagnóstico, Quito.

30

The Tagua Initiative: Building the Market for a Rain Forest Product

KAREN ZIFFER
Conservation International

FROM NUTS TO BUTTONS

Tagua buttons can once again be found on quality clothing in U.S. shops. Tagua, the nut of a palm tree that grows in the forests of northwestern South America, was once a very popular material for buttons and handicrafts. In the early part of the century, before cheap plastics were introduced, one of every five buttons manufactured in the United States was made of tagua. Tagua buttons remained popular in Europe for high fashion garment manufacturers because of its high quality and natural grain, but it virtually disappeared in the United States.

Now, through the Tagua Initiative, a broad range of tagua products, including buttons, jewelry, and carvings, are being brought to market to achieve grassroots economic development and rain forest conservation. Bearing the ™ mark, these products represent the unique collaboration between Ecuadoran tagua producers, local and international conservation groups, and forward-thinking U.S. businesses.

BIOLOGIC WEALTH AND ECONOMIC OPPORTUNITY

It has become evident that conservation cannot succeed in underdeveloped countries if it is contrary to the basic needs of local residents. Similarly, people cannot improve their economic condition if their natural resources are depleted and they lack access to income-generating activities. Poverty is a major driver of deforestation, and biologic wealth, properly managed, can offer economic opportunity.

The harvest and marketing of nontimber rain forest products is one way of promoting economic development in a manner consistent with wise resource use. Conservation International calls this strategy "conservation-based development." While there has been significant work done in the inventory of potentially useful species, there has been too little effort focused on building markets for these products in a manner that equitably benefits local producers and enhances the conservation of critical ecosystems. There continues to be a vast gap between the knowledge and tradition of forest residents and businesses eager to launch natural products that are ecologically sound. The Tagua Initiative is working to bridge that gap. The project has been designed by all of the participants to weave together the distinct strands of self-determined community development, applied scientific research, and good business.

Developing and introducing any product to the unforgiving marketplace is a difficult and often unsuccessful venture. The complications multiply when a number of social objectives are also being sought. Though rain forest products have become popular with consumers because of increasing awareness of the environment and demands for natural products, these products face stiff price competition from cheap synthetics, and they are often difficult to source reliably.

This chapter outlines briefly the mechanics of bringing tagua products to market. For a more detailed account of the community development aspect of the project, refer to Rodrigo Calero Hidalgo's chapter 29.

REVITALIZING THE TAGUA INDUSTRY

The Tagua Initiative is still in its infancy, but progress is being made in the communities, and the products are becoming commercially successful. The Tagua Initiative entered the U.S. market last year with the introduction of tagua buttons. In September of that year a CI press conference reached 100 million consumers through newspapers, magazines, television, and radio in the United States, Japan, Europe, and Latin America. Thousands of consumers are also learning about the Initiative—and about the importance of rain forest conservation—through clothing hangtags, in-store displays, and catalog descriptions. The first year of the program generated sales of 7 million buttons and nearly one-half million dollars of sales. New tagua products, including jewelry and carvings, are under development.

Step 1: Market Assessment

The feasibility of reintroducing tagua to the U.S. market began with investigating the history of the product, its traditional uses, the volumes exported over time, and the relationship between local producers, processing facilities, and end-consumers. An analysis of the current industry was also done to assess the possibility of growth and to determine whether there was a niche for a new commercial player with an environmental story to tell. This analysis suggested that CI could play a significant role in promoting tagua. This hypothesis was made concrete when CI was able to secure commitments from two U.S. clothing manufacturers who agreed to use tagua buttons if CI could ensure that their purchase would promote conservation. The participation by Patagonia and Smith and Hawken gave the project credibility, both in the eyes of the Ecuadorans and the U.S. business leaders with whom we were negotiating contracts.

Step 2: Design the Project with Local Partners

From the introduction of the TI concept, the local participants in the program have been integrally involved in the design and implementation of the project. Each partner has helped create the overall plan, even though each one has specific duties and responsibilities. Full local partnership is essential to the success of the project because conservation-based development is fundamentally rooted in the community, and because in business terms, the community and the local NGO are the production arm of the tagua business. If local partners are unaware of the demands imposed by international business markets, they cannot work to achieve these standards. Likewise, if the partners responsible for developing end-markets are not intimate with community constraints they are likely to make inappropriate demands and create unrealistic expectations.

Step 3: Structure Deals

On behalf of the project, Conservation International arranged agreements with business partners who wanted to manufacture TI tagua products. These licensing agreements were woven into the fabric of the project by requiring that 1) manufacturers purchase their raw materials from sources authorized by the project, in this way generating demand for the community-based enterprises and improving their bargaining position in relation to the processing factories, and 2) manufacturers pay a royalty fee based on tagua product sales. These fees help to fund the management and expansion of the project.

Step 4: Market, Market, Market

The success of the Tagua Initiative rests heavily on the ability to generate long-term demand for the material and organize equitable trade with the producers. To build demand the project team had to 1) develop a message: Tagua has been promoted as a high quality, beautiful, and versatile natural material. We have had to educate companies and consumers that tagua is an excellent replacement for animal ivory and is a renewable resource that can play a role in conserving tropical forests through sustainable community development, 2) develop a product strategy: The project began with buttons because tagua was traditionally used for buttons and there was established technology and an existing infrastructure that allowed the project to get started quickly. Target customers included outdoor clothing manufacturers, natural fiber companies, and high end designers. The market strategy needed to recognize the significant price difference between tagua and plastic buttons. We needed to find customers who were already using buttons from other natural materials (which are priced similarly to tagua) or who were willing to pay the difference.

Step 5: Product Expansion

The second phase of the project includes product diversification both of tagua and other forest resources from the community. This diversification is essential for the growth and stability of community's economic basis. Development is already underway to expand the tagua line into jewelry and handicrafts, and research is in progress to identify other commercially viable products.

Step 6: Value-Added at the Source

Building the local capacity to manage the tagua project and pursue its commercial opportunities has been central to the project since its inception. Rodrigo Calero Hidalgo's chapter 29 details the training program that CIDESA has implemented with the Commune membership to teach basic management techniques and to organize a system for taking greater control of the marketing of their resources. While the program began with organizing the collection and sale of the whole tagua nut, plans are underway to install some processing facilities using appropriate technology. In addition, an artisan training program is being designed to teach the creation and marketing of tagua handicrafts for the tourist market and for export.

A FEW REFLECTIONS

There is a demand for nontimber forest products. This demand is fueled by a number of trends. There is a growing awareness of the global environmental crisis and there is an increase in the desire for people to get involved through their purchasing decisions. In addition, the health and fitness trend is here to stay, which is reflected in the rapid growth of natural products markets. Finally, rain forest products are different; they appeal to consumer demands for the exotic and unique.

There are a growing number of businesses eager to use rain forest products, but they maintain tough sourcing standards. Almost every business of any size requires products with consistent quality, reliable supply, and competitive prices. In addition, they expect their suppliers to be professional, responsive, and in the case of rain forest marketing, deliver a credible and compelling story. It is difficult for new community-based projects to meet these standards immediately. Technical assistance and training are needed and flexibility needs to be built into the system.

Products should be carefully selected. First, they should be appropriate from a local perspective in terms of cultural traditions and ecological considerations. Second, they should be targeted for the correct markets, whether they be local, national, or international. Third, products should be selected that offer inherent quality and that have the potential for long-term demand. Products relying on the rain forest appeal as a fad will be short-lived to the detriment of producers who rely on them for their livelihood.

Business are wary of backlash. Too much preaching about the environmental message can scare businesses who fear they are inviting criticism and a thorough investigation of their entire operations. Claims should be honest and realistic.

A market assessment and strategy should be developed early on. Too many projects have failed in this arena because there has been insufficient attention to the commercial viability of a product marketing program. The conservation and development world is littered with projects that spent years identifying potential products and defining their biology when they ultimately could not be sold anywhere.

CONCLUSION

Providing the access to markets, and in this case access to international markets, remains a critical missing link in developing demand for sustainably harvested nontimber forest products. Quality products and a diversified strategy are needed to avoid the short-term boom and bust cycles that are devastating to local producers and often result in over-exploitation. But developing markets is difficult. It is made even more so by the imperative to develop them in a manner consistent with equitable community-based economic development and sound natural resource management. The challenge remains for the forces of development, conservation, and successful business to be reinforcing rather than contradictory.

31

Nontimber Forest Products from the Tropics: The European Perspective

MANFRED NIEKISCH
ORO VERDE: Foundation for the Conservation of Tropical Forests, Germany

Conservation of tropical forests is a key issue for conservation of biodiversity, and nontimber products may become an important key to conservation of tropical forests.

Sadly enough, we refer to "timber" to name such an incredibly big number and variety of animal and plant species and products thereof that we usually call "nontimber products." Therefore, a short look at the timber market might be useful to see what we can learn from the experiences with timber trade for the trade in nontimber products. The paltry terminology is an expression of the history of forest use and exploitation, seeing the forest at least primarily as a source of timber. All other "useful" elements and components of a forest ecosystem are simply "secondary forest products" or "nontimber products."

TROPICAL TIMBER IN EUROPE

Approximately 25 to 30 percent of all tropical timber (in terms of roundwood equivalent, RWE) exports was imported by countries of the European Community. That is about three times as much as was imported by the United States and close to the Japanese import figures. When measured in value, tropical hardwood imports into the European

Community exceeded the Japanese ones significantly (Enquete-Kommission, 1990; Nectoux and Dudley, 1987; Nectoux and Kuroda, 1989).

Now things in Europe are changing. Many conservation NGOs, the press, scientists, and—somewhat at the end of the line—the federal governments, have raised the issue of tropical timber and tropical forests. Alarmed by the disastrous destruction of tropical forests and its consequences, many European consumers are now willing to do their parts to halt this biodiversity holocaust. Many see only one way to help tropical forests: they refuse to buy any more tropical timber, since the timber trade is one of the main reasons for forest destruction.

This movement is growing stronger in many European countries. About 1,000 communities and cities in Germany have meanwhile decided not to use tropical timber in public buildings unless it comes from sustainably managed sources.

At the tenth meeting of the International Tropical Timber Council, ITTC, in June 1991 in Quito, Ecuador, the Vice Minister for the Environment of the German state of Berlin described the situation as follows:

> If the tropical countries cannot supply enough sustainably managed tropical timber, then they can be sure that not only the federal state of Berlin but also the whole of united Germany will be in a very short time no longer an important importer of tropical timber. I think that such a sharp decline of demand will happen in many other ecologically-minded consuming countries (Wicke, 1991).

The situation in the Netherlands is similar to that in Germany. Following campaigns of conservation NGOs and initiatives taken by numerous municipalities, the Dutch Government has come forward with an overall policy for the import and use (or nonuse) of tropical timber in the Netherlands.

The government paper presents a package of actions to assist producing countries in achieving sustainable forest management. Part of this package says that, from 1995 on, timber imports into the Netherlands will be restricted to producing countries with sustainably managed systems (Government of the Netherlands, 1991). A Dutch NGO has prepared a very successful brochure for communities on how to reduce the use of tropical timber (Sambeek, 1989).

In almost every European country there is a high demand from concerned consumers who are requesting information based on the guidelines established by *The Good Wood Guide* (Counsell, 1990), which lists alternatives to tropical timber.

ITTO's DILEMMA

According to a study for the International Tropical Timber Organization (ITTO), undertaken by the International Institute for Environment and Development (IIED) in 1988, not even 1 percent of the tropical timber in worldwide trade can be considered to come from sustainable sources (Poore et al., 1988).

Most of the recent exploitation of tropical forests for timber is simply overexploitation. If logging goes on at this current rate, at least 23 of the 33 countries that currently export wood will have to import.

The International Tropical Timber Council (ITTC), with its 44 member states, recognized the dilemma at its eighth meeting in May 1990 in Bali, Indonesia and declared that by the year 2000 all tropical timber in worldwide trade should come from "sustainable sources" only.

Unfortunately, the ITTC took no action to reach this target. At its tenth meeting, the ITTC turned down a proposal for concrete action to be taken on both the consumer's and producer's side and replaced the proposed action plan with a vague paper. The ITTC failed to assign any figure for the amount of forest area that must be brought under sustainable management. The ITTC failed as well to define prescriptions for what consuming countries ought to do for their share in achieving the so-called "target 2000." Consequently, it is most unlikely that this target can be reached in the remaining eight years.

Nevertheless, one decision of the tenth ITTC meeting may become historically important in the context of nontimber products. The ITTO will fund ($374,000, U.S.) a Brazilian project proposal (ITTO number PD 143/91) entitled "Nonwood Tropical Forest Products: Processing, Trade, and Collection."

The site for this 24-month project will be in Santarem in the Brazilian state of Para. In the same area, the Tapajos National Forest, ITTO has already supported the earlier stages of the implementation of a forest management and timber harvesting operation as a demonstration model area. With the incorporation of nonwood products, this could become an important integrated forest management project.

Such an integrated approach is a rare exception among all ITTO-projects. Still, it is interesting to observe that even a timber organization is starting to recognize the value of nontimber products. Nevertheless, be it timber or nontimber products, it is the question of sustainability that determines the difference between destruction and conservation of forest ecosystems.

SUSTAINABILITY

The concept of sustainable management has its origins in the temperate forest of Germany. The principle of sustained yield was formulated in 1713 by H. C. von Carlowitz, a German forester. This concept simply entails a level of harvest that can be maintained in perpetuity from a given forest area. Temperate forests are ecologically much simpler and better understood than tropical forests, and sustainable yield can be achieved relatively easily. Forestry and timber production in temperate forests have a long history and tradition and, contrary to tropical forests, there is a great deal of knowledge and broad experience available on which to draw.

Given the complexity of tropical forests, the general lack of knowledge about tropical silviculture, and the fact that the physical and cultural survival of many indigenous communities is dependent on intact forests, the expansion of "sustainable yield" to "sustainability," and its transfer from temperate to tropical forests pose two major problems. First, sustainability cannot be limited to the production and harvest of timber only. Second, given the lack of research and experience and the complexity of the ecosystem, there are great problems in determining quantities and qualities of sustained harvesting.

Management of tropical forests for any purpose, including the extraction of nontimber products, can be called sustainable only if it is sustainable at three levels. It has to be: 1) ecologically sustainable, 2) economically sustainable, and 3) socially sustainable. Otherwise any use will lead to the destruction of the ecosystem.

The plans for the Jatata project in the Bolivian Beni, managed by the Chimane Indians themselves, provides an excellent example for social sustainability. The reality of Jatata trade, as it has worked over the years until today, exploiting the Chimanes and violating their rights, is certainly socially unsustainable, but does (seen at the national level) seem to be economically sustainable. Somebody benefits—but it is the wrong people. This small example may be enough to explain the importance of social sustainability.

TIMBER IS MONEY

Everybody, including the conservation community, has to face the fact that timber from tropical forests is an important source of income for many developing countries and that those countries' economies depend (in some cases almost exclusively) on the export of tropical timber.

The prime advantage that tropical hardwood has on the international market compared to timber from temperate forest lies in its low price rather than its beauty, special physical properties, or other characteristics. The price for tropical timber is dictated by international trade companies, by the industrialized countries, and is heavily influenced by huge foreign debts. The producing countries are left with little—if any—choice. Therefore, one important step to help the developing countries' economies and ecologies would be to turn the high volume/low price trade in tropical timber into a low volume/high price trade. This is the responsibility of the importing countries.

NONTIMBER PRODUCTS: FOREIGN TRADE

Contrary to commercial timber exploitation, there are many examples of truly wise use of forests for nontimber products, in which the management is done by the local communities. Most of it up to now has been mainly for local consumption.

Extraction of nontimber products for international commercial utilization is likely to be possible in limited quantities only, and commercialization certainly carries risk of overexploitation, which would negatively affect biodiversity. But nontimber products from natural forests definitely have a better chance of being sustainably produced than tropical timber. A myriad of high value-added nontimber products (rather than raw material) and a strict high-price/low-volume trade have excellent potential for the sustainable economic use of many unaltered forest areas.

CHANGE THE MARKET STRUCTURES

At the moment, producing countries of tropical timber only receive a shameful 5 to 10 percent of the profits in timber trade, while 90 percent or more of the profits go to traders and others (based in the wealthy, industrialized nations) in the market chain. The last ones to receive benefit, if they receive any at all, are local communities and indigenous groups.

This unequal distribution of benefits, which is today's reality of trade in timber and in agricultural products, must be avoided in the future development of markets for nontimber products from tropical forests.

Otherwise, trade in these products will not meet the criteria of sustainability and will contribute to further destruction of the forests, just like the timber trade.

COICA (Coordinadora de las Organizaciones Indigenas de la Cuenca Amazonica), representing two million people of 300 indigenous groups, stated at the ITTO meeting in Ecuador that "there is no better way to sustain the forests than to recognize our land rights and our cultures— for the strengthening of our roots and the dreams of us all" (Nugkuag, pers. comm.).

In the same meeting, the Latin American NGOs declared in a joint statement: "Forest communities have the right to actively participate in the decision-making processes as well as in the execution of all activities involved in the establishment of sustainable management systems and to receive the benefits derived from these systems" (Organismos No Gubernamentales Latinosamericanos, 1991b).

Similar statements could be quoted from many other occasions, given by representatives of the Penan in Malaysia, as well as the Kayapo Indians of the Amazon, the Aeta in the Philippines, the Quechua Indians in the high Andes, and many others whose livelihood is seriously at risk because of imported human-made ecological disasters.

The outspoken wish of local communities and ethnic groups to be the initiators, managers, and the beneficiaries of projects in their environment has to be heard and respected by politicians and development aid institutions in what usually are called "donor countries."

Scientists and conservation organizations from the industrialized nations that like ORO VERDE receive constant in-put from and are in direct touch with the grassroots level in tropical countries, have to make sure that this message is heard and understood on the donor side. Environmental problems like the disastrous consequences of forest destruction are felt first and most directly at the grassroots level—to which government institutions and big companies normally have little, if any, access, and in which they may even lack interest.

SIGNS OF HOPE—OR JUST STRAW FIRES?

"Sustainable use," "buffer zone management," and "nontimber products" are already becoming buzzwords in international development aid. There is some indication that action is being taken to implement policies based on these principles. Consider the following examples from Germany: 1) The German Federal Ministry for Economic Cooperation (BMZ) is drafting a new forest-sector concept; nontimber products are likely to play a greater role than ever before. 2) In the past, traditional development aid was not able to halt the destruction of the environment—often it was the cause. Recognizing this and the need for new types of development aid, the BMZ has furthermore declared its strong political will to cooperate more closely with NGOs inside and outside the country: social sustainability is the desired outcome (Bundesministerium für Wirtschaftliche Zusammenarbeit, 1990). 3) The German development aid agency, Gesellschaft für Technische Zusammenarbeit (GTZ) organized, for the first time ever, in April 1991 an international workshop for its "people in the field" to analyze their experiences in nontimber products and to gather more knowledge on this issue. Follow-up activities (with regard to nontimber products) are underway. Furthermore, GTZ is preparing a new concept for nature conservation in technical cooperation (Esser, Ellenberg, and Niekisch, 1991).

These and other initiatives have been started because of countless NGO activities at all levels. They are partially the logical outcome of a hearing of many dozens of experts like foresters, ethnologists, and ecologists that was organized by a Study Commission of the German Parliament on the state of tropical forests. The 1,000-page report of this Commission is not only an invaluably rich source of information on tropical forests and all related problems, but gives also very clear recommendations to the German government (Enquete-Kommission, 1990).

There are similar tendencies in other European countries. It is up to NGOs and governments alike to push this process forward, carry it into other countries, and seize the opportunities resulting from it.

NONTIMBER PRODUCTS NEED CONSUMERS

North American and European consumers have proven over the last few years that they will respond to buyer-beware campaigns initiated by conservation NGOs. The examples of ivory, frog legs, spotted cats, and sea turtles amply demonstrate that consumers can change market structure.

Public awareness programs to decrease demand for certain products are certainly somewhat easier to plan than to establish new markets for nontimber products. This then will be a major new challenge for NGOs in Europe: to create markets for nontimber products from sources that are sustainable, and to link the producers as closely as possible to the markets.

If we wish to increase the number and role of nontimber products on the international market, then the basic requirements are consistency and continuity of production in terms of both quantity and quality; an international marketing system that creates and increases the demand for sustainably harvested products; a pricing system that makes the product attractive to the buyer; a marketing system that guarantees that a significant proportion of the income from sales goes back to the indigenous communities that hold the copyrights, intellectual property rights, and so on; a labeling system for sustainable products; and intensified research to find or develop new highly priced products (for example, perfume essences, medicines, and so on) for the international marketplace.

CONSERVATION THROUGH SOLIDARITY IN TRADE

If harvest and marketing of nontimber products from tropical forests are meant to be a contribution to conservation of natural ecosystems and of traditional culture (and not merely a new idea to line the pockets of Western companies), new forms of partnership between producers and consumers and new trade agreements will be required. Conservation of tropical forests must be the driving principle.

European consumers would be willing to purchase sustainably harvested products from the tropics, especially when buying them is a contribution to the conservation of tropical forests. During the last few years, this has been shown many times.

The analysis of market structures in the international timber trade is teaching us that trade in timber products from tropical forests should *not* be conducted. If we have learned this lesson and act accordingly, then destruction of millions of hectares of the tropical forests has not been in vain.

References

Bundesministerium für Wirtschaftliche Zusammenarbeit. 1990. *Hinweise für die Zusammenarbeit mit Nichtregierungsorganisationen der Entwicklungsländer in der staatlichen Entwicklungszusammenarbeit.* (Unpublished.)

Counsell, S. 1990. *The Good Wood Guide: A Friends of the Earth Handbook.* London: Friends of the Earth.

Enquete-Kommission "Vorsorge zum Schutz der Erdathmosphäre." 1990. *Zweiter Bericht. Schutz der tropischen Wälder.* Drucksache 11/7220. Deutscher Bundestag. Bonn.

Esser, J., L. Ellenberg, and M. Niekisch. 1991. *Technische Zusammenarbeit im Naturschutz. Überlegungen und empfehlungen für die GTZ, erstellt im Auftrag der Abt. 424.* Gesellschaft für Technische Zusammenarbeit. Eschborn (in print).

Government of the Netherlands. 1991. *Policy Paper on Tropical Rain Forests*. The Haag.

Nectoux, F., and N. Dudley. 1987. A Hard Wood Story. London: Friends of the Earth.

Nectoux, F., and Y. Kuroda. 1989. *Timber from the South Seas*. Gland, Switzerland: WWF.

Organismos No Gubernamentales Latinoamericanos. 1991a. *Pronunciamiento ante el consejo de la Organización Internacional de la Madera Tropical en su decimo perodo de sessiones*. (Unpublished.)

Poore, D. et al. 1988. *Natural Forest Management for Sustainable Timber Production*. Report for ITTO. International Institute for Environment and Development. London.

Sambeek, P. V. 1989. *Verminderen van het gebruijk van tropisch hout*. Amrsterdam: Vereiniging Milieudefensie.

Wicke, L. 1991. *Incentives for Sustainable Management of Tropical Forests*. (Unpublished.)

Wicke, L. 1991b. Pronunciamiento de los organismos no gubernamentales. Latinoamericanos ante el consejo de la organizacíon Internacional de la Madera Tropical en su decimo periodo de Sesiones. (Unpublished.)

32

Ethnobiology and U.S. Policy

KATY MORAN
Smithsonian Institution

As U.S. policy makers search for a more sustainable and equitable development process, ethnobiology can influence the discussion by demonstrating how commercialization of nontimber forest products provides an economic incentive to keep tropical forests intact. Likewise, it offers an opportunity for indigenous peoples who live in these forests to participate in and gain a more equitable share of the benefits of their country's development processes through their traditional knowledge of how to use and manage forest resources.

The search began because economic biases, particularly in multilateral development, had subordinated ecology and equity considerations. Natural scientists documented how development strategies that destroy forests by logging, mining, monoculture, cattle ranching, and huge hydroprojects erode the bilogical diversity of life itself. According to the World Resources Institute (1991), "We are committing an estimated 100 species a day to extinction, and the toll could rise to one fourth of all species over the next quarter century at current rates of habitat loss." Likewise, social scientists have documented disruption, and in extreme cases, decimation, of the diverse cultures of forest peoples that results from this type of development strategy. It causes the erosion of cultural diversity and loss of knowledge about the sustainable use of forest resources that its inhabitants have accumulated over millennia (Clay, 1986; Moran, 1987).

Diversity, both biological and cultural, is a measure of the number of types of components within a system, whether it be a natural system or a social system. It enhances the sustainability of a system because, when necessary, it supplies a greater range of options for adaptation. Diversity is valuable because, by spreading risk, it equips a system with alternatives for continuing survival if one of its components fails. As Western science would describe it, tropical forests represent laboratories of biologically diverse resources and tropical forest cultures represent similarly diverse libraries filled with information on how to use them. It is this complementary nexus of "biocultural diversity," within which ethnobiologists operate, that Western scientists now recognize as a critical but perishable resource as forests and forest cultures are destroyed.

Loss of biocultural diversity also has been recognized by policy makers who are initiating new policies to stem its loss and to capitalize on its use. This began on October 17, 1986, when the U.S. House of Representatives Committee on Science, Space, and Technology and its subcommittee on Natural Resources, Agriculture Research, and Environment requested a study from the Office of Technology Assessment (OTA) on how international aid agencies could better match development technologies to local environmental conditions of recipient countries. To use natural resources technologies can have a high or a low level impact on the environment, depending on the type and the volume of resources the technology requires and the amount and kind of waste it produces. For example, traditional technology to harvest nontimber forest products typically is diversified and resource-extensive. It does not deplete, nor concentrate pressure on a single forest resource, but distributes human impacts across the larger forest ecosystem (Warren, 1989).

Conclusions of the OTA study were not surprising—the use of ecologically appropriate technologies is essential if developing countries are to stabilize their fragile economies. Damage to their resource base can be critical because less-developed countries ultimately depend more directly on natural resources within their borders. The report also expressed the need to support local participation in the development process through institution building in developing countries. It recommended aid be channeled specifically toward strengthening the role and capacity of existing indigenous resource managers who can identify viable alternative development approaches (OTA Staff Paper, 1987).

The report generated belief that the current development approach, particularly as demonstrated by multilateral forestry projects, was too narrowly focused on an increase in economic productivity as an indicator of success. It did not accurately reflect costs such as loss of ecological services or social impacts of forestry projects. Profits from trees, mainly for the timber they provide, skewed and subordinated ecological considerations such as species loss, diminishment of the forest capacity to absorb increasing carbon dioxide emissions, exacerbation of soil erosion and flooding, increased siltation in rivers and estuaries, impairment of watersheds, and disruption of hydrological cycles of large regions.

Likewise, logging roads have social costs as they open huge tracts of forest to the landless poor fleeing from urban shanty towns. Urban colonists attempt to practice a version of shifting cultivation that is not sustainable with high population levels or constricted fallow periods and results in one of the leading causes of tropical deforestation. These poorest of the poor, as well as indigenous peoples, suffer first and worst from deforestation because they live in and depend on tropical forests as their only resource. According to Matthews (1989), "Traced through its effects on agriculture, energy supply and water resources, tropical deforestation impoverishes about a billion people."

Too often, logging also results in capitalism that has now become institutionalized in many tropical countries. The local political processes are corrupted as elites are bribed and rewarded with logging concessions for political allegiance. It undermines, weakens fragile new democracies, exacerbates social divisions, and exposes the blatant disregard of equity considerations in development strategies that focus only on economic productivity. The OTA report recommendation for local participation in development illuminated the necessity of a more equitable distribution of development benefits. Equity, or a measure of how evenly benefits are distributed among recipients, was recognized as an essential component, or the "third e" of a balanced development process that incorporates economy, ecology, and equity (World Resources Institute, 1991).

THE ROLE OF ETHNOBIOLOGY IN U.S. DEVELOPMENT POLICY

From ethnobiological data, there is growing recognition that besides timber, forests contain a cornucopia of food, fodder, fiber, fuel, and other forest products that are economically viable, ecologically sustainable, and equitably beneficial. A recent study of a small patch of Amazon rain forest found that the total net revenue generated by the sustainable extraction of "minor" forest products are two to three times higher than those resulting from a one-time exploitative use such as timber (see Gentry, chapter 15). Over time, fruits and latex represented 90 percent of the potential value of the forest, and timber only 10 percent (Peters et al., 1989). The same is true for unevaluated species found in tropical forests regarding drug development. Although some 7,000 natural compounds are used in modern medicines, fewer than 2 percent of higher plants have been thoroughly screened for biological activity. Yet in 1985, Principe (1989) found that almost half of the world market value of drugs, estimated at $90 billion (U.S. dollars), came from medicinal plants.

There are both benefits and constraints to the development paradigm of commercialization of nontimber forest products. Benefits are those of a balanced development process that incorporates economy, ecology, and equity. The forest's wealth of raw materials can be sustainably harvested to generate foreign trade income at the national level, and to create employment and income to forest peoples at the local level, preserving the forest as an intact socioeconomic and ecological entity.

One constraint is tapping the vital source of indigenous peoples' knowledge that Western science has only recently begun to understand and value. Over millennia, diverse groups in tropical forests have developed varied cultural regimes to utilize, maintain, and express knowledge of their resources. Embedded in complex cultural systems and expressed through local languages, symbols, rituals, songs, and narratives, these systematic bodies of knowledge categorize and describe the relationships between cultures and their environment. However, ethnobiological methods can generate data about the value and uses for tropical plant and animal resources and technologies to manage them (Frake, 1962; Sturtevant, 1964; Rappaport, 1968; Vayda, 1969).

More importantly, ethnobiologists operate with indigenous peoples at two levels of confluence that require, in return, two sets of responsibilities. At one level we elicit and analyze traditional knowledge, then correlate it with Western science. Research responsibilities that this convergence entails are now developing as professional codes from ethics committees of various ethnobiolgical societies (Boom, 1990). At the same time, ethnobiologists bridge cultures of the industrialized and the nonindustrialized world. As small scale cultures become linked to a market economy, processes are set in motion that will modify their social organization to accommodate this change. Ethnobiologists who speak the language and live and work with small scale cultures can help mitigate and manage modifications in community dynamics that inevitably accompany this level of convergence. It will be crucial to use harvesting and marketing techniques that are controlled by local people and that unify communities through use of traditional or compatible institutions (Colchester, 1989).

Greater constraints exist in opposing the powerful groups with vested interests in maintaining their profitable timber operations. Since indigenous peoples who live in tropical forests are nearly powerless and politically, geographically, and linguistically vulnerable, ethnobiologists often have become advocates for their empowerment. Many ethnobiologists have played catalytic roles in some of the following initiatives for U.S. policies on indigenous peoples and their natural resources (Kloppenburg, 1987; Moran, 1989; Boom, 1990; Posey, 1990; King, 1991).

U.S. POLICIES ON INDIGENOUS PEOPLES AND NATURAL RESOURCES

Forestry Convention

During the General Assembly of the United Nations in 1989, it was agreed to hold a United Nations Conference on Environment and Development (UNCED) in June 1992 in Rio de Janeiro, Brazil. UNCED participants will discuss a wide range of issues, including indigenous peoples and their participation in economic development.

The United States has placed forest management and protection as a priority of UNCED, and is currently developing specific objectives for a binding international convention on forest management and conservation for signature in Rio. Although there is opposition from many countries to a legally binding forest convention, the idea of a statement of principle has been widely endorsed. Discussions by U.S. government agencies and nongovernmental organizations have resulted in agreement that a basic principle be included on the needs of indigenous peoples who use the forests as the basis for their livelihood, social organization, or cultural identity, and who have an economic stake in sustainable forest use. Actions to achieve this include promoting means for indigenous peoples and members of local communities to participate in decision-making processes for proposed forest-related actions where their interests are potentially affected. Another proposed action is to encourage the exchange and preservation of technical, scientific, and traditional knowledge between Western scientists and indigenous peoples.

Discussions were also held on how, as democracy increasingly becomes recognized as an organizing principle for political systems throughout the world, recognition of the rights of indigenous peoples to play a leading role in shaping their own destiny can be more explicit (U.S. Agency for International Development, 1990). Proposed bilateral actions focused on institution-building as the key to effective participation in policy dialogue and national decision-making. This could provide the foundations for a broad range of development activities, such as marketing cooperatives for nontimber tropical forest products and microenterprise for development. Other proposed bilateral actions are to identify ways to enhance the value of standing forests through 1) policy reform that more accurately reflects the true costs and benefits of alternative forestry activites, 2) identification of economically valuable forest species, including timber and nontimber flora and fauna, and 3) development of improved and sustainable extraction schemes for both. Other discussions on proposed points the United States should promote in both governmental and nongovernmental forums included 1) collecting a synthesis of social, biophysical, and institutional initiatives that combine indigenous and external scientific knowledge, 2) promoting tenure security to safeguard sustainable common property resource management regimes, and 3) activities to support meaningful dialogues between indigenous peoples and their governments, including enhancement of local abilities to articulate appropriate political demands.

S. 748 Pan-American Cultural Survival Act of 1991

Despite the trend toward democracy and constitutional reform in Latin America over the last 15 years, Senator Alan Cranston (D-Calif.) points out that 30 to 40 million indigenous peoples of Latin America still remain at the margins of political democracy. Consequently, Senator Cranston, a member of the Senate Foreign Relations Committee, introduced S. 748 to assist indigenous peoples of Central and South America to take meaningful and representative roles in their nation's democratic institutions and practices, as well as to assist them in protecting their lands, cultures, and natural resources. The bill calls for a comprehensive review of the rights of indigenous peoples in each country of Latin America, as well as an assessment of indigenous peoples' participation in national democratic institutions and practices.

Advocates lobbied to include a part of Senator Cranston's bill in reauthorization of the U.S. Foreign Assistance Act (FAA). Report language accompanied the Senate version of the FAA reauthorization, which stated that "maintaining commercial proprietorship of the traditional knowledge of plant and animal resources gives indigenous peoples a stronger voice in how their natural resources are managed" (Anderson, 1991).

Additionally, Senator Cranston's bill generated an issue brief by the Congressional Research Service to inform members of Congress about current concerns and activities of indigenous peoples in Latin America. "Latin American Indigenous Peoples and Considerations for U.S. Foreign Assistance" states that in recent years, analysts have recognized a rapid and dynamic growth of political salience and efficacy by indigenous groups (Serafino et al., 1991). These groups have demanded increased participation in their political system and greater levels of autonomy and self-determination for indigenous communities. Their purpose is to preserve their lands, cultures, and languages, and to secure greater socioeconomic rights and benefits. The brief contains a historic overview and demographic information, then suggests options for the use of U.S. foreign aid to empower indigenous peoples and increase their level of political participation in the current systems.

House Congressional Resolution 354

Maintaining commercial proprietorship for traditional knowledge of plant and animal resources by indigenous peoples was also addressed in the U.S. House of Representatives by Representative John Porter. He founded and co-chairs the Congressional Human Rights Caucus, and also has an assignment to the Committee on Appropriations Subcommittee on Foreign Operations, Export Financing, and Related Programs in the House. Since the subcommittee appropriates funding for U.S. bilateral and multilateral aid, Porter is in a position to influence the development process and has established a reputation for responsible initiatives that promote sustainable development actions.

Representative Porter framed commercial proprietorship of traditional knowledge as an intellectual property rights issue for international trade discussions, when, on July 23, 1990, he introduced House Congressional Resolution 354. The resolution asked the U.S. trade representative to discontinue negotiations at the Uruguay round of the ongoing General Agreement on Tariffs and Trade (GATT) talks until discussions of intellectual property rights (IPR) were extended to include protection for the traditional knowledge indigenous peoples have of their plant and animal resources.

Disregarding consideration of traditional knowledge in trade discussions, other IPR issues in international trade talks are fraught with complexity and contention. IPR discussions have been further complicated by biotechnology issues that are beyond the scope of this chapter (Lewis, 1990). Nevertheless, Representative Porter believed that it was in this arena that the issue should be addressed because an urgent side of the GATT debate had been neglected by negotiators due to lack of an existing consistent or inclusive international program to monitor ownership of, or protection for, traditional knowledge. His goal was to compensate indigenous informants, through contractual agreements, for plant selection for laboratory scrutiny or for information on the pharmacological activity of a given species. It is more difficult to patent a plant–derived drug than a synthetically or microbally derived one, but it can be done. It requires demonstrating a novel compound and use, patenting the process by which it is made, or patenting a chemically modified version of the compound.

Others besides Porter have stated the value of traditional knowledge, especially for leads to research on pharmaceuticals.

> The accomplishments of aboriginal people in learning plant properties must be a result of a long and intimate association with, and utter dependence on, their ambient vegetation. This native knowledge warrants careful and critical attention on the part of modern scientific efforts. If phytochemists must randomly investigate the constituents of biological effects of 80,000 species of Amazon plants, the task may never be finished. Concentrating first on those species that people have lived and experimented with for millennia offers a short-cut to the discovery of new medically or industrially useful compounds. (Schultes, 1988)

For example, information on 119 known useful plant-derived drugs was analyzed to determine how many were discovered due to traditional knowledge on the plants from which they were isolated. Analysis of plant-based products on the market has shown that 74 percent have the same or related use in Western medicine as originally used by indigenous curers (Farnsworth, 1988). If selected randomly, estimations are that one in 10,000 to 35,000 plant samples will yield a medically useful activity. Yet as many as 25 percent of the prescriptions in the United States contain natural products extracted from plants. In a recent collection in Belize, Balick (1990) found 25 percent of the ethnobotany collection showed preliminary activity, compared to 6 percent of active plants in the random collection. Including herbals and medicines sold over-the-counter, the estimated value of plant-based drugs was $11 billion in 1985 in the United States (Principe, 1989). Porter felt that if traditional knowledge is the productive basis for selecting plants for research strategies and laboratory analysis, it is a genuine claim for compensation.

By failing to acknowledge and place value on this knowledge, Porter feared the United States would overlook a critical opportunity for sustainable development. It could promote the conservation of biological resources in situ by ecosystems, sustain the livelihoods and lifestyles of indigenous cultures which choose to continue that method of subsistence, and distribute benefits to the technicians who had discovered, maintained, and innovated this knowledge within their cultures for generations.

House Congressional Resolution 354 directed the U.S. trade representative to undertake a study of the social, environmental, and economic implications of IPR trade policies for biological resources, and to assess prior traditional uses for such resources by indigenous peoples. The measure failed to pass during the 101st Congress, but nevertheless, raised consciousness of the bioethics of ethnobiology research and generated international discussions and national actions related to intellectual property rights for traditional knowledge.

UNITED STATES AGENCY FOR INTERNATIONAL DEVELOPMENT (USAID)

The U.S. Agency for International Development (USAID) is currently developing guidelines in the form of a project annex to ensure that intellectual property created or transferred during the course of any USAID-funded activity is adequately and effectively protected. This annex for intellectual property rights, now in draft form, is seen as the basic mechanism through which the United States and host country governments jointly agree to protect intellectual property created or furnished under a USAID project or program.

The rationale of USAID for ensuring the equitable distribution of rights is that adequate protection of intellectual property is necessary to promote innovation that is a critical element to establish free and competitive markets. The strategy for USAID to promote sustainable economic growth in developing countries must therefore ensure that inventors of intellectual property are legally protected and fairly compensated.

Within the USAID program, ethnobiology carries a critical responsibility. For example through the "Useful Plants and Traditional Healing Systems in Belize" project, USAID collaborates with the New York Botanical Garden to study the biocultural diversity of Belize. Project components include efforts to publish information on Maya medical systems, including documentation of the plants used in herbal medicine, a discussion of the effectiveness of these plants within their cultural context, as well as results from pharmaceutical screening programs. Emphasis is on strengthening ties between Belezian and U.S. scientists regarding preservation of biocultural diversity.

The project has developed a collaborative network of healers that includes both generalists and specialists in the hierarchy of traditional medicine. The rich cultural diversity of Belize is reflected in the number of people from different backgrounds included in this network of Maya, Ladino, Creole, East Indian, and Mennonite. Each culture focuses on different disease concepts and the use of an independent but convergent suite of plants (Schweitzer, 1991).

NATIONAL INSTITUTE OF HEALTH/NATIONAL SCIENCE FOUNDATION

Many pharmacological or chemical studies reported in scientific journals begin by quoting the traditional use of a given plant in developing countries of the tropics where both the traditional use of the plant and the plant material itself were collected (Elizabetsky, 1990). Tropical species have yielded important medical compounds including vincristine, vinblastine, reserpine, tubocurarine, diosgenin, pilocarpine, and quinine.

Recognizing the contribution of ethnobiology, the National Institutes of Health and the National Science Foundation are collaborating with USAID to develop a joint program in drug discovery, biodiversity conservation, and economic growth. The project will promote the conservation of biodiversity through the discovery of bioactive agents from natural products. An emphasis of the project is to ensure that equitable economic benefits from these discoveries accrue to the country of origin, and the appropriate institutions within that country.

Ethnobiology will be a critical component of the new effort, with a focus based on several important premises. First, traditional knowledge is as threatened and as valuable as biodiversity. Second, traditional healers have proprietary knowledge of local plant and animal resources, as well as observational and analytical skills from which that knowledge is derived and expanded, which must be afforded appropriate protection and adequate compensation. Third, the maintenance of biological diversity and traditional knowledge must make sense economically within the cultural, political, and developmental realities in countries where these resources occur. The joint NIH, NSF, USAID project also recognizes the need to educate both decision makers and the general public to assure broad and effective recognition of the value of biocultural diversity for conservation and sustainable development.

CONCLUSION

As demonstrated by the above U.S. policy initiatives, the role of ethnobiology can be important in many ways.

1. Ethnobiological data demonstrate and give value to ways indige-

nous communities have conducted research and development on their local flora and fauna for drugs, agricultural techniques, natural resource management, and other uses.

2. As is likewise essential for the policy process, practicing ethnobiology requires information, methods, and cooperation from many interdisciplinary perspectives. Anthropology, botany, zoology, pharmacology, chemistry, economics, ecology, and agriculture are some disciplines involved in ethnobiology. It draws on and merges the intellectual capital of both the natural and social sciences as each discipline adds its particular component to the field and to the policy process.

3. The interdisciplinary perspective of ethnobiology helps to frame and answer larger policy initiatives, such as methodologies for sustainable development, within the bounds of a single discipline. It integrates shared, cross-disciplinary concerns and challenges, and offers an opportunity to establish a broad-based constituency for collaborative actions to promote common goals and values.

4. Locales for ethnobiological studies have historically been in tropical ecosystems of the most endangered flora and fauna. These are the fragile environments where most marginal and minority cultures of the world in greatest danger of extinction are located. Field ethnobiologists who learn their languages and cultures are able to articulate concerns of these often voiceless people to policy makers.

Ethnobiology is the nexus for the natural and social sciences, traditional and modern knowledge, and North and South relationships. It demonstrates the partnership of the natural and social sciences in the Brundtland Commission's definition of sustainable development: "The means by which development is made to meet the needs of the present without compromising the ability of future generations to meet their own needs." This definition implies a social contract both within and between generations that total biological assets will not be reduced in the long run as a result of the development process.

These linkages within ethnobiology demonstrate that it is impossible to isolate economic, social, and environmental problems. They help us understand how most environmental degradation that we witness today represents symptoms of unresolved socioeconomic problems such as an overdependence on resource-intensive and pollution-prone technologies, destructive population and consumption patterns, and inequity among and within nations. Policy solutions to these global problems require recognition of their relationship to each other. Ethnobiology can continue to play an important role in that process.

References

Anderson, M. 1991. *Memorandum: Report Language Suggestions for the U.S. Foreign Assistance Act.* Committee on Foreign Relations, United States Senate.

Balick, M. J. 1990. Ethnobotany and the Identification of Therapeutic Agents from the Rain Forest. *Bioactive Compounds from Plants: Proceedings of the Ciba Foundation Symposium 154.* New York: Wiley & Sons.

Boom, B. M. 1990. Giving Native People a Share of the Profits. *Garden Bol.* 14 (6):28–31.

Clay, J. (ed.). 1986. Multilateral Banks and Indigenous Peoples: Development or Destruction? *Cultural Survival Quarterly* 10 (1):1–80.

Colchester, M. 1989. Indian Development in Amazonia: Risks and Strategies. *The Ecologist,* 19 (6):249–54.

Colchester, M. 1991. *Forest Peoples and Sustainable Forest Management.* Paper presented at World Resources Institute Colloquium on Sustainability in Natural Tropical Forest Management. Washington, D.C.: World Resources Institute.

Elizabetsky, E., and Z. C. Castillos. 1990. Plants Used as Analgesics by Amazonian Cabados as a Basis for Selecting Plants for Investigation. *International Journal for Crude Drug Research* 28:49–60.

Farnsworth, N. R. 1988. Screening Plants for New Medicines. In *Biodiversity* (E.O. Wilson, ed.). Washington, D.C.: National Academy Press.

Frake, C. 1962. Cultural Ecology and Ethnography. American Anthropologist, Part I. 64 (1)53–59.

King, S. R. 1991. The Source of Our Cures. *Cultural Survival Quarterly.* Summer: 19–22.

Kloppenburg, J., Jr. 1987. *First Seed: The Political Economy of Plant Biotechnology, 1492–1900.* New York: Cambridge University Press.

Lewis, P. 1990. United States Urged to Join Effort to Preserve Plants' Gene Resources. *New York Times,* April 24, 1990.

Matthews, J. Tuchman. 1989. Redefining Security. *Foreign Affairs,* Winter: 165.

Moran, K. 1987. Traditional Elephant Management in Sri Lanka. *Cultural Survival Quarterly: Grassroots Economic Development.* 11 (1):23–26.

Moran, K. 1989. Washington Watch. *Practicing Anthropology* 11 (2):1–2.

Office of Technolgoy Assessment (OTA) Staff Paper. 1987. *Aid to Developing Countries: The Technology/Ecology Fit.* OTA, Congress of the United States. Washington D.C.

Peters, C. M., A. H. Gentry, and R. O. Mendelsohn. 1989. Valuation of an Amazonian Rainforest. *Nature* (339):655–56.

Posey, D. 1990. Intellectual Property Rights and Just Compensation for Indigenous Knowledge. *Anthropology Today* 6 (4):13–16.

Principe, P. 1989. The Economics and Significance of Plants and their Constituents as Drugs. In *Economic and Medicinal Plant Research,* H. Wagner, H. Hikino, and N. R. Farnsworth (eds.). New York: Academic Press.

Rappaport, R. 1968. *Pigs for the Ancestors.* New Haven, CT: Yale University Press.

Schultes, R.E. 1988. Letter to the Editor. Tradition, Conservation, and Development: Occasional Newsletter of the Commission on Ecology's Working Group on Traditional Ecological Knowledge. October: 10.

Serafino, Nina M. et al. 1991. *Latin American Indigenous Peoples and Considerations for U.S. Assistance.* Washington D.C., Congressional Research Service, The Library of Congress.

Sturtevant, William C. 1964. Studies in Ethnoscience. *American Anthropologist* 66 (3):91–131.

United States Agency for International Development. 1990. *The Democracy Initiative.* Washington, D.C.: USAID Bureau for Program and Policy Coordination.

Vayda, A. P., and R. Rappaport. 1969. *Environment and Cultural Behavior.* New York: Natural History Press.

Warren, M. D. (ed.) 1989. *Indigenous Knowledge Systems: Implications for Agriculture and International Development.* Ames, Iowa: Iowa State University Technologies and Social Change Program.

World Resources Institute. 1991. *World Resources Institute Colloquium on Sustainability in Natural Forest Management.* World Resources Institute. Washington, D.C.

World Resources Institute. 1991. The Transition to a Sustainable Society. *WRI Issues and Ideas.* Washington, D.C.: World Resources Institute.

33

Some General Principles and Strategies for Developing Markets in North America and Europe for Nontimber Forest Products

JASON CLAY
Cultural Survival

Over the past few years, we have learned that the most successful strategies for rain forest conservation are ones that both maintain the land's natural biodiversity and that meet the economic needs of forest peoples. Scientists are now coming to understand what forest residents have long known—namely, that standing forests are theoretically capable of generating more income and employment than the same areas cleared for pasture or agriculture. Yet a practical model of sustainable development with more equitable relations with external markets has never been fully implemented or tested.

For this reason, in 1989 Cultural Survival established Cultural Survival Enterprises (CSE) to create and test an alternative income-generation model based on marketing nontimber rain forest products. CSE is working with rain forest communities throughout the world that are trying to expand their income through sustainable extractive activities. By developing and expanding markets for nontimber forest products—products that are harvested without destroying forest—we hope to provide a compelling economic incentive to local rain forest producers, First and Third World governments, NGOs, international banking organizations, and development agencies to demonstrate to all concerned that rain forests are best protected by providing forest residents with sustainable sources of income.

During our first year of operation (September 1, 1989 to August 31, 1990) we imported nontimber commodities and sold them at a billing price of $349,000 to 17 companies which in turn used them to manufacture 19 products. Another 75 companies are doing research and development for future product launches in the coming one to three years.

For the past year, we have been able to sell the *idea* of marketing rain forest products with a logically argued theory rather than facts to back up the case. Based on the first year of trade we are now able to give concrete information and analyses to back up the theories. That information is included in the 11 lessons (detailed below) that we have learned during the first full year of trade. Trade during the second year will apparently increase about eightfold, slightly down from our projection because we were unable to borrow the $1 million of working capital in time to finance the purchase of commodities that we wanted to trade this year.

Here are the most important lessons that we have learned in the past year, lessons that we feel should be studied by any and all wishing to enter the field of trade as a means of supporting local conservation and development. While most of these lessons are derived from our experiences with international trade, we believe that they would be applicable, with modification, to local, regional, or national trade as well.

1. Start With What's Already On the Market. Marketing efforts should focus initially on products for which there are already markets and therefore substantial production that can be redirected to new products/customers. Existing products offer the best chance of quickly creating and benefiting from international markets. Even so the lag time will be from five to ten years for significant impact. For example, most new products take from one and a half to three years to develop to the point of manufacture. It will take a year for sales to grow to significant levels and any type of profit sharing or licensing agreements would only begin to be received up to six months after the end of the first profitable year.

It will be much more difficult to bring new products into either U.S. or European markets. In fact, European markets are likely to become even more difficult to penetrate after the 1992 economic unification. In the meantime, health and safety data from controlled experiments and, usually, animal testing, are required before any raw material can be included in food or cosmetics. It is not enough to argue that people in the Amazon have been using the product for centuries. Such testing, combined with product development, in the pharmaceutical industry takes an average of 20 to 25 years. Food and cosmetic industries would require somewhat less time.

2. Diversify Product and Reduce Dependence On a Few Products. The diversification of products being sold is absolutely essential to the overall viability of extractivism. However, such diversification must be created one product at a time and will take decades to include a number of products in the overall market mix, much less include a number of different producer groups and organizations.

The strategies of Brazilian forest peoples should focus, initially, on Brazil nuts, rubber, and babassu. Much can be done to increase the value received from these items, even though it will not happen within the next year or so. Many groups do not have these commodities, however. Therefore, other extractivists could focus on acai, cupuassu, copaiba, andiroba, urucum, or cashews. In each region the high volume (and/or value) items that most people depend on for a substantial portion of their income should be used to create the markets, market mechanisms, production facilities, momentum, and financing that it will take to bring on other, lesser known commodities.

In choosing new priority products, attention should be paid to the value generated per unit of labor, as well as the complementability of the product in question to the subsistence base and other income generating activities of the group(s) in question.

3. *Diversify the Markets for Raw and Processed Forest Products.* In order to reduce the overall risk to producer groups, attempts should be made to diversify the number and type of end users of nontimber forest products. This needs to be done for each product. For example, Brazil nuts can be used as nuts (shelled or in the shell), ice cream, baked goods/ confections, cereal, candy, oil, flour, and so on. Risk can be further reduced by penetrating regular markets, as well as markets for organic wild or natural products. Finally, each commodity could well have local, regional, national, *and* international markets.

4. *Add Value Locally.* Examine the marketing mechanisms of each product to determine the best ways to capture the value that is always added to a product as it leaves the source. In some cases, this could be accomplished by simply transporting the product further into the market and eliminating other traders. In other cases, value could be added by processing the commodity locally, as in the case of Brazil nuts or rubber. Each attempt to add value can in fact often lead to a doubling of the income from the product. In addition, each attempt to add value can also have unexpected side benefits. For example, shelling Brazil nuts in the houses of natives not only increases the value of the nuts they collect, it also reduces the overall costs by reducing by two-thirds of the weight and volume of the nuts and thus their transportation costs. In fact, forest shelling of nuts for those who care to undertake such activities allows many groups to sell nuts on the market for the first time because they can now make money at it.

In addition, the local organizations and market mechanisms that allow groups to sell *onto* the market as groups, also allow them to purchase *from* the market in volume as groups, thereby reducing their overall costs for consumer necessities.

5. Capture Value That Is Added Further From the Source. All markets in the world add value to products with every transaction or transfer, as well as with increased scarcity or distance from the source of production. The value added is somewhat related to labor and capital investment, but it is equally as related to scarcity and monopoly. As a rule of thumb, however, greater values are added with each transaction that takes the product further from the source and with each process that transforms or alters the form of the product.

For example, trucking Brazil nuts from Xapuri into Rio Branco might double the price (and therefore should be done, if local processing cannot be undertaken), but charging 5 percent on Brazil nuts sold in New York City would actually generate more money per pound than doubling the price locally. Likewise, the cost of wholesale nuts in New York City only represents about 5 percent of the *gross* sales of a candy manufacturer using 25 percent nuts by weight in his or her product. Therefore, a 5 percent premium on the sale of nuts to the manufacturer generates only about half as much revenue as a 5 percent profit sharing agreement. These are not hard and fast rules because the profitability of companies varies a great deal. However, the principles hold. Producers, or their supporters, should work to capture some of the value that is added to each step in the commercial system until the product reaches the final consumer.

This does not imply, however, that the goal should be to go after profits at one point in the system versus another, but rather that there are many points in the system where revenues can be generated that can be returned to the original producers. This money can be used in the short term to help groups organize to protect their land rights, evaluate and adapt their resource management strategies, and, perhaps eventually, establish new marketing and/or processing systems.

Whatever mechanisms are established to add value, the producer groups that are selling into the system should receive 100 percent of the value that is added to their production. For example, if a group is providing 10 percent of a company's demand, they should receive 10 percent of any and all returned revenues (for example, profit sharing). The other 90 percent could be used to protect the basic human and land rights of forest groups in the country of origin of the raw materials, help finance their organizing efforts, determine the most appropriate adaptation and monitoring of the use of resources, and finally, identify and help other local groups begin to process their own produce so that they can take advantage of revenues that can be generated in the market-place.

How the remaining 90 percent of the revenue is to be used could be determined by a committee composed of representatives of the various forest peoples' organizations, as well as the Brazilian human rights and environmental organizations that work closely with them. For funds resulting from Cultural Survival's marketing efforts, CS would want one representative on such a group to ensure that funds were being expended in line with our nonprofit status and objectives and in keeping with the spirit and legal agreements in which they are being raised.

6. *Proposed Solutions Must Equal the Scope of the Problems.* The destruction of rain forests and the displacement of rain forest peoples is proceeding at unprecedented rates. Solutions to such problems must be large enough to eventually affect all forests and all forest peoples. For that reason, it is not wise to see as solutions any hand-picked or pet projects that are highly subsidized and therefore not replicable at the scale that would be necessary to solve the problems of many communities.

This being said, solutions do have to be divisible. They must start with one village, group, or forest at a time and build from there. It is also the case that no single solution will solve all the problems of the rain forest or of forest peoples. Still, the overall lesson holds; we cannot apply tiny bandages to problems that clearly require tourniquets.

7. *No Single Forest Group Can Provide Enough Commodities for Even a Small Company in North America or Europe.* The production of a single group in the Amazon is not sufficient to meet the needs of even a small company in North America or Europe. Large companies are out of the question. We spoke with a large candy company about the possibility of using rain forest nuts in a candy bar. They use 70 metric tons of nuts per eight-hour production shift, a *year's* production of the Xapuri nut shelling plant.

If forest peoples want to market produce under their own logo or mark, they will have to aim at small markets where overheads will be large and net profits returned to the groups small. In addition, because the costs per unit of selling smaller quantities are larger than those for larger quantities, the end price to the consumer is higher. This means that the products will not be used by poor or middle class people and any possibility of creating broad-based consumer support for rain forest peoples will be lost.

In general, more money can be earned by charging 5 percent on raw materials and receiving even 1 percent of profit sharing from a company using 500 metric tons of nuts than getting 100 percent of the net profits on the sale of 25 metric tons of nuts.

8. Controlling a Large Market Share of a Commodity Allows Considerable Influence Over the Entire Market. Controlling a large share of the market for single commodities can be used to lobby for the return of value added and profits generated to forest residents' organizations or those organizations that work closely with them. However, this can be accomplished only if all groups work together closely.

CSE customers, for example, now use a sizable percentage of the Brazil nut market in the United States. With this collective market share, CSE can begin to leverage control over other sales. If a broker wants to supply our customers, then we can argue that they will have to charge their customers an environmental premium on all Brazil nut sales, with 100 percent of this premium being returned to forest residents' organizations or their support groups. These premiums would be used for "conservation"—land rights, political organizing, resource management, and processing and marketing of forest products.

Advantages of large volume trade can also be used to exert influence over cashew nuts, babassu, copaiba, and achiote.

Nor is the influence of an honest trader on an economic system limited to international trade. In 1991, when Xapuri calculated what it could pay collectors for nuts in the shell and still operate the factory profitably, the figure was twice that of the going rate in Acre. Within two weeks, collectors throughout the state were demanding, and in most cases receiving, higher prices for their nuts.

9. Make a Decent Profit in the Marketplace, Not a Killing. Do not expect to make a killing in the marketplace—if it were so simple, many more people would be doing it. However it is essential and possible that value be added locally and overall income increased. Likewise, each decision to do away with an intermediary or middle person must be made with a thorough understanding of all the functions that are served by that person before he or she is eliminated. Somebody has to step in to provide the goods and services. Even though these systems are clearly exploitative, unless other organizations are prepared to take them over, local producers could actually be worse off if existing systems are destroyed and not replaced.

How value can best be added will vary a great deal with each product. Research must be done on collecting, transporting, processing, and marketing before decisions are made about which areas are best entered first, second, or not at all. Factors that affect these decisions are management skills, transportation equipment, warehouse space, and working capital.

In most cases, it is best, in the short term at least, to leave some if not most of the existing structure in place until it is more thoroughly understood. Again, what exactly is changed and when will vary by product and region, as well as by the producer group and its social relations of production with the market.

10. The Markets in North America and Europe Are for Saving the Rain Forests and for Conservation Rather Than for Forest People. It is a sad fact that the awareness of rain forest issues outside of those countries where they exist is limited to a concern for plants and animals. Few potential consumers of such products even know that the forests are occupied. They do not understand intuitively that raw materials are produced by people. The growing green consumer movement in North America and Europe for nontimber rain forest products represents an attempt to get beyond concerns about recyclable packaging and personal health issues (for example, organic versus pesticide-laden) so that consumers can also become aware that their purchases can help to protect living environments and the peoples who have in the past and continue to use and maintain them.

11. Certification of Environmental Sustainability Is Key. Because the growing green market in the United States and Europe is primarily for conservation, the sale of commodities must be linked with systems to ensure that the quantity of products being taken from the forest does not destroy the very resource that consumers are spending money to help protect.

As an aside, the harvest of any product from a forest will probably affect in some way the future diversity of that forest. However, the manipulation of forests has long taken place, even in those areas that are today considered pristine. For that reason, we do not want to say that the products being harvested and sold will not change the forest, because they most certainly will.

What we need, then, are careful environmental impact assessments and monitoring systems that examine from the beginning of increased trade onward the impact of harvesting rates on reproduction of the species in the forest. (Remember, I am talking about items already for sale on the market; for new products the assessments would have to be undertaken first.) Such systems need to be worked out for each commodity over time, with the involvement of both local communities and scientists. In the end, however, it is the local communities whose continued markets (and livelihoods) would depend on such certification.

CONCLUSION

These are the major lessons that have been learned in the first year. It is clear that each of these will be further refined or even refuted over time. Also, other lessons will be learned as it is almost certain that the longer we engage in these activities, the more commonly held assumptions will be challenged, and some at least will be abandoned as a result of new insights.

These current insights are offered to help forest peoples think about their own strategies. We would be happy to provide any additional information as it may be required.

Conclusions and Recommendations

A large and growing body of ethnobotanical knowledge demonstrates the incredible variety of botanical resources capable of supplying the material, nutritional, medicinal, and spiritual benefits to human societies. To date, this knowledge has been insufficiently applied to the process of socioeconomic development, and the intellectual contribution of indigenous cultures has been insufficiently recognized and inadequately compensated.

This traditional ethnobotanical knowledge of indigenous peoples and other long-term forest dwellers lays the cornerstone for a new form of socioeconomic development in tropical rain forests. However, the value of this knowledge will be lost if the following issues are not addressed during the identification and commercialization of rain forest products:

1. Forest dwellers must be assured of land and resource tenure, and their knowledge of forest plants must be accorded the status of intellectual property.

2. Trade restrictions and regulatory barriers to the introduction of tropical products in international markets must be eased.

3. The ecology of tropical forest plants (seasonal abundance, pollination and dispersal biology, regeneration rates, and so on) must be reflected in strategies for harvesting and marketing new products.

4. Indigenous and local communities must be direct participants in the research, planning, and decision making processes involved in the development of nontimber forest products.

5. The development needs and management capacities of forest communities must be respected and enhanced gradually, to minimize the risk of disruption posed by rapid economic change.

6. Businesses in importing countries must subscribe to a code of ethics that includes financial compensation to originating communities, such as payment of royalties.

7. Conservationists must increase their efforts to support the self-determination of indigenous forest dwellers, to link product development to existing and newly initiated conservation programs in critical rain forest areas, and to raise public awareness of the importance of nontimber forest products in both importing and exporting countries.

To hasten the expansion of viable markets guided by ecological and cultural sensitivity, and to return economic benefits to rain forest communities in time to halt, and ultimately reverse, the deforestation of the tropics, the participants of the Panama City workshop directed the following individuals and institutions to enact specific recommendations:

Ethnobotanists and biologists documenting the biodiversity and cultural uses of tropical rain forests:

• To acknowledge the intellectual contribution of forest peoples to the expansion of scientific knowledge of tropical biodiversity
• To assist their collaborators in tropical rain forests to gain recognition for their knowledge and receive equitable compensation for the use of their intellectual property
• To ensure that knowledge in the form of environmental education, publications, monographs, bibliographic materials, and so on is repatriated to the countries, regions, and communities in which fieldwork is carried out

Governments and bilateral and multilateral development aid agencies:

• To support the development of national and local institutions dedicated to documenting tropical biodiversity and ethnobotanical knowledge, and applying this knowledge in local and regional development activities
• To reevaluate the economic viability of current development activities in tropical rain forest areas and to assess the potential of nontimber forest products in present and planned development projects
• To revise national and international policies that stand in the way of commercialization of rain forest products and limit the flow of economic benefits to local communities
• To cooperate with conservation groups and commercial enterprises to encourage investment in nontimber forest products and manage these investments in ways that limit ecological disruption and maximize social benefits

Nonprofit conservation and cultural advocacy organizations:

• To seek opportunities to blend the commercialization of rain forest products with present and planned conservation activities in tropical rain forests

• To pursue collaborative joint ventures with commercial enterprises and organizations representing forest dwellers

• To acknowledge the irreplaceable contribution of indigenous peoples to our knowledge of tropical ecosystems, and to advance the concept that this knowledge represents intellectual property deserving equitable compensation

• To support product commercialization projects with public awareness activities that increase public understanding of the ecological and ethical issues at stake in managing tropical forests sustainably

For-profit businesses and commercial enterprises:

• To support fair and equitable arrangements to compensate forest dwellers and harvesters, both for materials harvested from tropical rain forests and for the cultural knowledge that leads to the identification of potential products

• To submit voluntarily to ecological review by tropical ecologists of proposed marketing of tropical rain forest plant materials

• To participate in defining ethical guidelines for the harvesting, processing, and marketing of tropical rain forest products

• To use product marketing campaigns to educate consumers about rain forest products as well as to enhance brand recognition and increase sales.

• To support the socioeconomic development of communities where the raw materials of new rain forest products originate

The participants at the Panama City workshop share a belief that the sustainable harvesting and marketing of rain forest products offer a unique opportunity to blend conservation with viable community development.

Index

About the Authors

Mr. Jorge Añez
c/o Guillermo Rioja
Ave. Villazon 1958
Edificio Villazon
Piso 10–Ept. 10A
La Paz, BOLIVIA

Dr. Rodrigo G. Bernal
Instituto de Ciencias Naturales
Universidad Nacional de Colombia
Apartado 7495
Bogotá, COLOMBIA

Dr. Antonio Brack
Bolognesi
159 2nd Piso
Miraflores
Lima, PERU

Mr. Geodisio Castillo
Fundación Dobbo Yala
Apartado Postal 83–0308
Zona 3
Panama, Rep. de PANAMA

Dr. Jason Clay
Cultural Survival
53-A Church St.
Cambridge, MA 02138

Dr. Douglas Daly
The New York Botanical Garden
Bronx, NY 10460

Dr. Robert DeFilipps
Department of Botany
Smithsonian Institution
Washington, D.C. 20560

Dr. James Duke
U.S. Department of Agriculture, ARS
Building 001, Rm. 133 BARC-West
Beltsville, MD 20705

Dr. Alwyn Gentry
Missouri Botanical Garden
Box 299
St. Louis, Missouri 63166–0299

Dr. Pierre Grenand
ORSTOM
213 rue La Fayette
75480 Paris, FRANCE

Dr. Mahabir P. Gupta/Dr. Pablo Solís
Universidad De Panama
FLORPAN
Apartado Postal 10767
Estafeta Universitaria
Republica de PANAMA

Dr. Rodrigo Calero Hidalgo
CIDESA
Casilla 608
Sucursal 12 Octubre
Quito, ECUADOR

Dr. Luz Graciela Joly
Universidad Nacional de Panama
David, Rep. de PANAMA

Dr. Steven R. King
Shaman Pharmaceuticals, Inc.
887 Industrial Road, Suite G
San Carlos, CA 94070–3312

Dr. Sonia Lagos-Witte
Jl. Siaga Raya Kav. 3
Rt. 014/003
Pejaten Barat–Pasar Minggu
Jakarta, 12510 INDONESIA

Ms. Constanza La Rotta
Universidad Nacional Autonoma de
Mexico
A.P. 6–807
Col. Juarez Del Cuauhtemoc
C.P. 06600, Mexico D.F.
MEXICO

Dr. Jean-Paul Lescure
ORSTOM/INPA
Caixa Postal 478
69011 Manaus
BRAZIL

Mr. Gary J. Martin
University of California–Berkeley
94 Boulevard Flandrin
75116 Paris
FRANCE

Mr. Peter H. May
UN Food and Agricultural
Organization
Via delle Terme di Caracalla
00100 Rome, ITALY

Mr. Juan Mayr
Fundación Pro-Sierra Nevada de Santa
Marta
Calle 74 No.2–86
Piso 2
Bogotá, COLOMBIA

Ms. Katy Moran
S.I. Building "Castle"
Smithsonian Institution
1000 Jefferson Drive, SW Room 230
Washington, D.C. 20560

Dr. Scott Mori
The New York Botanical Garden
Bronx, NY 10460

Dr. Gary Paul Nabhan
Native Seed/SEARCH
Rt. 1 Box 100
Ajo, AZ 85321

Dr. Manfred Niekisch
ORO VERDE
Bodenstedtstrabe 4
D–6000 Frankfurt/Main 70
GERMANY

Mr. Henrik B. Pedersen
Botanical Institute
University of Aarhus
Nordlandsvs
68,8240 Risskov, DENMARK

Mr. Linwood H. Pendelton
Harvard University
102 Line St., #2
Somerville, MA 02143

Dr. Darrell A. Posey
Museu Goeldi - Brazil
Zoologische Staatssmalung
Munchenhhaussenstrasse 21
D–800
GERMANY

Mr. Conrad Reining
Conservation International
1015 18th Street, NW
Suite 1000
Washington, D.C. 20036

Mr. Guillermo Rioja
Conservation International
Casilla–5633
Ave. Villazon 1958
Edificio Villazon
Piso 10–Ept. 10A
La Paz, BOLIVIA

Dr. Jan Salick
303 Porter Hall
Ohio University
Dept. of Botany
Athens, OH 45701–2979

Dr. Richard E. Schultes
Botanical Museum of Harvard
University
Harvard University
Cambridge, MA 02138

Dr. Victor M. Toledo
Centro de Ecologia
Universidad Nacional Autónoma de
Mexico
Apartado Postal 70–275
Mexico, DF–04510
MEXICO

Karen Ziffer
Conservation International
1015 18th Street, NW
Suite 1000
Washington, D.C. 20036

Also Available
from Island Press

For a complete catalog of Island Press publications, please write: Island Press, Box 7, Covelo, CA 95428, or call: 1–800–828–1302.